Research and Development on a Salt Processing Alternative for High-Level Waste at the Savannah River Site

Committee on Radionuclide Separation Processes for High-Level Waste at the Savannah River Site

Board on Radioactive Waste Management
Board on Chemical Sciences and Technology
Division on Earth and Life Studies

National Research Council

NATIONAL ACADEMY PRESS
Washington, D.C.

NOTICE: The project that is the subject of this report was approved by the Governing Board of the National Research Council, whose members are drawn from the councils of the National Academy of Sciences, the National Academy of Engineering, and the Institute of Medicine. The members of the committee responsible for the report were chosen for their special competence and with regard for appropriate balance.

Support for this study was provided by the U.S. Department of Energy, under Grant No. DE-FC01-99EW59049. All opinions, findings, conclusions, and recommendations expressed herein are those of the authors and do not necessarily reflect the views of the Department of Energy.

International Standard Book Number: 0-309-07593-9

Additional copies of this report are available from:

National Academy Press
2101 Constitution Avenue, N.W.
Box 285
Washington, DC 20055
800-624-6242
202-334-3313 (in the Washington metropolitan area)
http://www.nas.edu

Copyright 2001 by the National Academy of Sciences. All rights reserved.

Printed in the United States of America.

THE NATIONAL ACADEMIES
Advisers to the Nation on Science, Engineering, and Medicine

National Academy of Sciences
National Academy of Engineering
Institute of Medicine
National Research Council

The **National Academy of Sciences** is a private, nonprofit, self-perpetuating society of distinguished scholars engaged in scientific and engineering research, dedicated to the furtherance of science and technology and to their use for the general welfare. Upon the authority of the charter granted to it by the Congress in 1863, the Academy has a mandate that requires it to advise the federal government on scientific and technical matters. Dr. Bruce Alberts is president of the National Academy of Sciences.

The **National Academy of Engineering** was established in 1964, under the charter of the National Academy of Sciences, as a parallel organization of outstanding engineers. It is autonomous in its administration and in the selection of its members, sharing with the National Academy of Sciences the responsibility for advising the federal government. The National Academy of Engineering also sponsors engineering programs aimed at meeting national needs, encourages education and research, and recognizes the superior achievements of engineers. Dr. William A. Wulf is president of the National Academy of Engineering.

The **Institute of Medicine** was established in 1970 by the National Academy of Sciences to secure the services of eminent members of appropriate professions in the examination of policy matters pertaining to the health of the public. The Institute acts under the responsibility given to the National Academy of Sciences by its congressional charter to be an adviser to the federal government and, upon its own initiative, to identify issues of medical care, research, and education. Dr. Kenneth Shine is president of the Institute of Medicine.

The **National Research Council** was organized by the National Academy of Sciences in 1916 to associate the broad community of science and technology with the Academy's purposes of furthering knowledge and of advising the federal government. Functioning in accordance with general policies determined by the Academy, the Council has become the principal operating agency of both the National Academy of Sciences and the National Academy of Engineering in providing services to the government, the public, and the scientific and engineering communities. The Council is administered jointly by both Academies and the Institute of Medicine. Dr. Bruce Alberts and Dr. William A. Wulf are chairman and vice-chairman, respectively, of the National Research Council.

COMMITTEE ON RADIONULIDE SEPARATION PROCESSES FOR HIGH-LEVEL WASTE AT THE SAVANNAH RIVER SITE

MILTON LEVENSON, *Chair*, Bechtel International (retired), Menlo Park, California
GREGORY R. CHOPPIN, *Vice-Chair,* Florida State University, Tallahassee
JOHN E. BERCAW, California Institute of Technology, Pasadena
DARYLE H. BUSCH, University of Kansas, Lawrence
JAMES H. ESPENSON, Iowa State University, Ames
GEORGE E. KELLER II, Union Carbide Corporation (retired), South Charleston, West Virginia
THEODORE A. KOCH, E.I. Du Pont De Nemours And Company (retired), Wilmington, Delaware
ALFRED P. SATTELBERGER, Los Alamos National Laboratory, Los Alamos, New Mexico
MARTIN J. STEINDLER, Argonne National Laboratory (retired), Downers Grove, Illinois

Staff

ROBERT S. ANDREWS, Study Director, Board on Radioactive Waste Management (through January 2001)
CHRISTOPHER K. MURPHY, Study Director, Board on Chemical Sciences and Technology
TONI GREENLEAF, Administrative Associate, Board on Radioactive Waste Management
LAURA D. LLANOS, Senior Project Assistant, Board on Radioactive Waste Management
ANGELA R. TAYLOR, Senior Project Assistant, Board on Radioactive Waste Management

BOARD ON RADIOACTIVE WASTE MANAGEMENT

JOHN F. AHEARNE, *Chair*, Sigma Xi and Duke University, Research Triangle Park, North Carolina
CHARLES MCCOMBIE, *Vice-Chair*, Consultant, Gipf-Oberfrick, Switzerland
ROBERT M. BERNERO, U.S. Nuclear Regulatory Commission (retired), Gaithersburg, Maryland
ROBERT J. BUDNITZ, Future Resources Associates, Inc., Berkeley, California
GREGORY R. CHOPPIN, Florida State University, Tallahassee
RODNEY EWING, University of Michigan, Ann Arbor
JAMES H. JOHNSON, JR., Howard University, Washington, D.C.
ROGER E. KASPERSON, Stockholm Environment Institute, Stockholm, Sweden
NIKOLAY LAVEROV, Russian Academy of Sciences, Moscow
JANE C. S. LONG, University of Nevada, Reno
ALEXANDER MACLACHLAN, E.I. Du Pont de Nemours & Company (retired), Wilmington, Delaware
WILLIAM A. MILLS, Oak Ridge Associated Universities (retired), Olney, Maryland
MARTIN J. STEINDLER, Argonne National Laboratory (retired), Downers Grove, Illinois
ATSUYUKI SUZUKI, University of Tokyo, Japan
JOHN J. TAYLOR, Electric Power Research Institute (retired), Palo Alto, California
VICTORIA J. TSCHINKEL, Landers and Parsons, Tallahassee, Florida

Staff

KEVIN D. CROWLEY, Director
MICAH D. LOWENTHAL, Staff Officer
BARBARA PASTINA, Staff Officer
GREGORY H. SYMMES, Senior Staff Officer
JOHN R. WILEY, Senior Staff Officer
SUSAN B. MOCKLER, Research Associate
TONI GREENLEAF, Administrative Associate
DARLA J. THOMPSON, Senior Project Assistant/Research Assistant
LATRICIA C. BAILEY, Senior Project Assistant
LAURA D. LLANOS, Senior Project Assistant
ANGELA R. TAYLOR, Senior Project Assistant
JAMES YATES, JR., Office Assistant

BOARD ON CHEMICAL SCIENCES AND TECHNOLOGY

KENNETH N. RAYMOND, *Co-Chair*, University of California, Berkeley
JOHN L. ANDERSON, *Co-Chair*, Carnegie Mellon University, Pittsburgh, Pennsylvania
JOSEPH M. DESIMONE, University of North Carolina and North Carolina State University, Raleigh
CATHERINE C. FENSELAU, University of Maryland, College Park
ALICE P. GAST, Stanford University, Stanford, California
RICHARD M. GROSS, Dow Chemical Company, Midland, Michigan
NANCY B. JACKSON, Sandia National Laboratory, Albuquerque, New Mexico
GEORGE E. KELLER II, Union Carbide Company (retired), South Charleston, West Virginia
SANGTAE KIM, Eli Lilly and Company, Indianapolis, Indiana
WILLIAM KLEMPERER, Harvard University, Cambridge, Massachusetts
THOMAS J. MEYER, Los Alamos National Laboratory, Los Alamos, New Mexico
PAUL J. REIDER, Merck Research Laboratories, Rahway, New Jersey
LYNN F. SCHNEEMEYER, Bell Laboratories, Murray Hill, New Jersey
MARTIN B. SHERWIN, ChemVen Group, Inc., Boca Raton, Florida
JEFFREY J. SIIROLA, Chemical Process Research Laboratory, Kingsport, Tennessee
CHRISTINE S. SLOANE, General Motors, Troy, Michigan
ARNOLD F. STANCELL, Georgia Institute of Technology, Atlanta
PETER J. STANG, University of Utah, Salt Lake City
JOHN C. TULLY, Yale University, New Haven, Connecticutt
CHI-HUEY WONG, Scripps Research Institute, La Jolla, California
STEVEN W. YATES, University of Kentucky, Lexington

Staff

DOUGLAS J. RABER, Director
RUTH MCDIARMID, Program Officer
CHRISTOPHER K. MURPHY, Program Officer
SYBIL A. PAIGE, Administrative Associate

PREFACE

The committee held three meetings and completed two reports in just over six months, a feat that would not have been possible without the assistance of many individuals and organizations. The committee received excellent support from Department of Energy, Westinghouse Savannah River Company, and staff from several national laboratories during the course of this study. On behalf of the committee, I want to acknowledge and thank Kenneth Lang (U.S. Department of Energy) and Harry Harmon (Pacific Northwest National Laboratory), who served as the committee's main points of contact and helped to organize the presentations at the committee's three information-gathering meetings. I also want to thank Jerry Morin and Joe Carter (Westinghouse Savannah River Company) for their help in unraveling the complexities of the high-level waste system at Savannah River, and the other individuals listed in Appendix D who provided briefings to the committee during its information-gathering meetings in Washington, D.C. and in Augusta, Georgia.

The completion of this study also would not have been possible without the support of the Board on Radioactive Waste Management (BRWM) and Board on Chemical Sciences and Technology (BCST). On behalf of the committee, I especially want to acknowledge and thank study directors Robert Andrews (BRWM) and Chris Murphy (BCST), board directors Kevin Crowley (BRWM) and Doug Raber (BCST), and BRWM senior project assistants Laura Llanos, Toni Greenleaf, and Angela Taylor.

Finally, I want to acknowledge and thank my colleagues on the committee, all of whom spent an unusual amount of their time over the past six months preparing for and attending committee meetings and generating and reviewing report drafts. It was a pleasure to lead such a capable group, and I hope that our collective efforts have contributed in some small way to helping the nation address its defense waste legacy in a responsible and cost-effective manner.

Milton Levenson, Chair
June 2001

LIST OF REVIEWERS

This report has been reviewed in draft form by individuals chosen for their diverse perspectives and technical expertise, in accordance with procedures approved by the NRC's Report Review Committee. The purpose of this independent review is to provide candid and critical comments that will assist the institution in making its published report as sound as possible and to ensure that the report meets institutional standards for objectivity, evidence, and responsiveness to the study charge. The review comments and draft manuscript remain confidential to protect the integrity of the deliberative process. We wish to thank the following individuals for their review of this report:

J. Brent Hiskey, University of Arizona
Edward Lahoda, Westinghouse Science and Technology Department
Kenneth N. Raymond, University of California, Berkeley
Lanny Robbins, Dow Chemical Company
Della Roy, Pennsylvania State University
Stephen Yates, University of Kentucky
Edwin L. Zebroski, Elgis Consulting

Although the reviewers listed above have provided many constructive comments and suggestions, they were not asked to endorse the conclusions or recommendations nor did they see the final draft of the report before its release. The review of this report was overseen by Royce W. Murray, University of North Carolina, appointed by the National Research Council, who was responsible for making certain that an independent examination of this report was carried out in accordance with institutional procedures and that all review comments were carefully considered. Responsibility for the final content of this report rests entirely with the authoring committee and the institution.

CONTENTS

EXECUTIVE SUMMARY, **1**

INTRODUCTION, **5**

PROGRESS AND RESULTS OF DOE'S RESEARCH AND DEVELOPMENT PROGRAM, **8**
Small Tank Tetraphenylborate Precipitation, **9**
Crystalline Silicotitanate Ion Exchange, **15**
Caustic Side Solvent Extraction, **22**
Actinide and Strontium Removal, **28**

REFERENCES, **34**

APPENDIXES
Appendix A Letters of Request for this Study, **35**
Appendix B Interim Report, **40**
Appendix C Biographical Sketches of Committee Members, **77**
Appendix D Information-Gathering Meetings, **81**
Appendix E Acronyms and Abbreviations, **84**

EXECUTIVE SUMMARY

The U.S. Department of Energy (DOE) is nearing a decision on how to process 30 million gallons of high-level radioactive waste salt solutions at the Savannah River Site in South Carolina to remove strontium, actinides, and cesium for immobilization in glass and eventual shipment to a geologic repository. The department is sponsoring research and development (R&D)[1] work on four alternative processes and plans to use the results to make a downselection decision in a June 2001 time frame. The DOE requested that the National Research Council help inform this decision by addressing the following charge:

1. evaluate the adequacy of the criteria that will be used by the department to select from among the candidate processes under consideration;
2. evaluate the progress and results of the research and development work that is being undertaken on these candidate processes; and
3. assess whether the technical uncertainties have been sufficiently resolved to proceed with downsizing the list of candidate processes.

The committee's interim report (Appendix B) served as a response to the first point of this charge. In that report, the committee found that DOE's proposed criteria are an acceptable basis for selecting among the candidate processes under consideration, but that the criteria should not be implemented in a way that relies on a single numerical "total score." Responses to the last two points are provided in this report.

A previous National Research Council report (NRC, 2000) found that there were potential barriers to implementation of all of the alternative processing options. A recommendation was made that Savannah River should proceed with a carefully planned and managed research and development program until enough information is available to make a defensible and transparent downselection decision. As a result of this report, DOE has developed and is vigorously pursuing an R&D program to resolve several outstanding issues related to the selection and implementation of these alternative processes. Consequently, technical,

---

[1] A list of acronyms and abbreviations may be found in Appendix E.

schedule, and cost risks have been more clearly defined and in some cases significantly reduced. The present committee has evaluated this R&D work and offers the following evaluations.

RESPONSE TO CHARGE 2: EVALUATE THE PROGRESS AND RESULTS OF THE RESEARCH AND DEVELOPMENT WORK THAT IS BEING UNDERTAKEN ON THESE CANDIDATE PROCESSES

The committee has provided a detailed evaluation of the R&D work in the second section of this report. On the basis of this review, the committee has identified the following unresolved issues for each alternative process:

Small Tank Tetraphenylborate Precipitation (STTP)
- The process by which tetraphenylborate (TPB) decomposes is not completely understood and *is not predictable*, either mechanistically or empirically.
- An unresolved issue is how cesium removal can be accomplished with waste batches where unexpected TPB decomposition occurs so rapidly that expected decontamination factors (DFs) are not achieved.
- Final selection of an antifoam agent has not been made.

Crystalline Silicotitanate (CST) Nonelutable Ion Exchange
- The mechanism of aluminosilicate precipitation on CST is not yet understood. This issue poses a potentially high technical risk.
- The reliance on a single supplier for CST poses a high schedule risk.
- Technical uncertainties—including column plugging, resistance to hydraulic transfer, irreversible desorption, and column system technologies—will continue to constitute a high risk for the use of this process for cesium removal. The method and resources required to resolve these risks are not clear.

Caustic Side Solvent Extraction (CSSX)
- Successful bench-scale demonstration of the complete CSSX process with actual tank waste is critical. These demonstrations are needed to clarify any residual risks.

Actinide and Strontium Removal: Monosodium Titanate (MST)

- Two alternate precipitation processes are competitive with MST and should be studied further. These employ sodium nonatitanate, which behaves similarly to MST, and sodium permanganate. The selection of a process for actinide and strontium removal is mostly independent of the selection of a cesium removal process.

CHARGE 3: ASSESS WHETHER THE TECHNICAL UNCERTAINTIES HAVE BEEN SUFFICIENTLY RESOLVED TO PROCEED WITH DOWNSIZING THE LIST OF CANDIDATE PROCESSES

This report has focused exclusively on technical issues related to the candidate processes for radionuclide removal from high-level waste salt solutions at Savannah River. However, because the final downselection must be based on a number of issues in addition to science and technology, the committee makes no recommendation on which process(es) should be selected. Rather, the committee has attempted to identify residual technical risks that should be a component in the decision-making process for downselecting the list of candidate processes.

The committee believes, however, that technical uncertainties have been resolved sufficiently to proceed with downselecting the list of candidate processes. To this end, the committee offers the following advice:

Small Tank Tetraphenylborate Precipitation

- The STTP process has remaining technical uncertainties, but engineering solutions for most of these problems probably can be found. However, because of the unpredictability of the decomposition rate of the TPB, there remains the risk that one or more of the 67 high-level waste production batches will require additional or special treatment before it can be processed using this option.

Crystalline Silicotitanate Nonelutable Ion Exchange

- Of the three cesium separation processes under consideration, it is the committee's judgment that CST has the most technical uncertainties and the highest technical risks.

Caustic Side Solvent Extraction
- Unless tests with actual waste encounter new problems, the CSSX option for cesium separation presents, at present, the fewest technical uncertainties of any of the three cesium separation alternatives.

Actinide and Strontium Removal: Monosodium Titanate
- All of the cesium separation processes depend upon a separate step to remove strontium, neptunium, and plutonium. Currently, that step uses MST. Because the success of this step is essential to all three of the processes for cesium separation, the committee believes that continued R&D on an alternate process to MST for removal of actinides and strontium is essential until MST processing can be demonstrated to meet the saltstone, Defense Waste Processing Facility (DWPF) throughput, and DWPF glass requirements.

INTRODUCTION

The U.S. Department of Energy (DOE) is nearing a decision on how to process 30 million gallons of high-level radioactive waste salt solutions[1] at the Savannah River Site (SRS) in South Carolina to remove strontium, actinides, and cesium for immobilization in glass and eventual shipment to a geologic repository. The department is sponsoring research and development (R&D) work on four alternative processes at Savannah River and at several national laboratories, and it plans to use the results to make a downselection decision in a June 2001 time frame. This decision will involve the selection of one or more alternatives for further R&D, pilot plant demonstration, and near the end of this decade, implementation to initiate processing of high-level waste at the site.

The department has requested technical advice from the National Research Council (NRC) to inform this downselection decision. In 1999, the DOE requested advice from the NRC on its efforts to identify potential processing alternatives and assess their technical feasibility. A National Research Council committee, hereafter referred to as the "2000 NRC committee," was formed to undertake that work and issued its advice in an interim and final report (NRC, 1999, 2000, respectively). These reports raised numerous concerns about the processing alternatives and identified important issues to be addressed by the department's R&D program. The department's current R&D efforts are focused on addressing the issues identified in these reports.

Following the issuance of that final report, the DOE requested that the National Research Council continue to advise the department on the downselection process (Appendix A). The present committee was formed to undertake this work. It comprises seven members of the first committee and two new members (Appendix C). The committee was given the following charge:

- evaluate the adequacy of the criteria that will be used by the department to select from among the candidate processes under consideration;

---

[1] The waste to be processed includes supernate and salt cake that is stored in the underground high-level waste tanks at the site. The sludge portion of the waste is now being removed from the tanks, washed, and sent to the Defense Waste Processing Facility to be immobilized in glass.

- evaluate the progress and results of the research and development work that is being undertaken on these candidate processes; and
- assess whether the technical uncertainties have been sufficiently resolved to proceed with downsizing the list of candidate processes.

The department also invited the committee, at its discretion, to provide comments on the implementation of the selected process.

This report focuses exclusively on the technical issues related to the candidate processes for radionuclide removal from high-level waste salt solutions at SRS. However, because final downselection must be based on a number of issues in addition to science and technology, the committee does not believe it is appropriate to recommend which process(es) should be selected. Rather, the committee has attempted to identify residual technical risks that should be a component of the decision-making process for downselecting the list of candidate processes. Some of these risks are a normal part of practice and scale up, while other risks encountered in the R&D program may reflect potential problems with the use of these process for radionuclide separation.

The present committee issued an interim report in March 2001 that addresses the first point of its charge. That report is reproduced in Appendix B and can be viewed on-line at http://www.nap.edu. It is not discussed further in this final report. The present report addresses the second and third points and the discretionary component of the charge.

The committee recognizes that the primary audience for this report is the assistant secretary for environmental management, who requested this study, her management team, other high-level DOE managers, and Congress. Therefore, the committee has striven to be concise rather than comprehensive and has intentionally avoided inclusion of the voluminous introductory and background sections that are a characteristic of many National Research Council reports. The final report of the 2000 NRC committee (NRC, 2000) provides an excellent summary of the high-level waste program at Savannah River, the candidate processing options, and the department's R&D program. That report is available for sale from National Academy Press and can be viewed on-line at http://www.nap.edu. Detailed technical background on cesium separation at the Savannah River Site and associated laboratories may be found in the DOE report *Savannah River Site Processing Project Research and Development Plan* (Pacific Northwest National Laboratory, 2000).

The information used to develop this report was obtained from briefings provided to the committee by the department and its contractors at the Savannah River Site, Oak Ridge National Laboratory (ORNL), and Argonne National Laboratory (ANL) at committee meetings in February

Introduction 7

and March 2001. A list of presentations received by the committee is provided in Appendix D.

The presentations to the committee were generally excellent. However, given the accelerated schedule for the R&D program and this review, complete, fully documented results generally were not available for the committee's review. The committee did, however, have an opportunity to discuss R&D results with the project scientists who attended its meetings. Therefore, the committee has used its best collective judgment in evaluating these results and formulating its findings and recommendations, but wishes to acknowledge that its evaluation is necessarily incomplete because of these information limitations. Similarly, given the tight schedule for this review, the committee has not performed an evaluation on how extensive any additional research and development needs to be. Instead, the committee has sought to identify those areas of technical risk that warrant further investigation.

PROGRESS AND RESULTS OF DOE'S RESEARCH AND DEVELOPMENT PROGRAM

This section provides an evaluation of DOE's R&D program on the three alternative processes for the removal of cesium and one alternative process for the removal of strontium and actinides from the high-level waste at SRS. The committee has used the conclusions and recommendations from the final report of the 2000 NRC committee (NRC, 2000) as a starting point in its evaluation and has reviewed the R&D results generated since that report was issued to determine if the issues raised by that committee have been addressed adequately.

This section is organized as follows: For each alternative process, a brief recapitulation of the relevant conclusions and recommendations of the 2000 NRC committee is provided. This is followed by an analysis of current R&D efforts under way at SRS, other national laboratories, and academic laboratories to address the 2000 NRC committee's findings and recommendations. The present committee presents findings that identify any technical uncertainties that it believes warrant further consideration and, where appropriate, makes recommendations to address them. In addition, the committee provides its conclusions on the state of resolution of technical uncertainties and its impact on the downselection process.

The four processes under primary consideration by DOE include one process for removal of strontium and actinides from high-level waste with a nonelutable ion exchange process utilizing one or more sodium titanate compounds and three candidate processes for cesium removal: caustic side solvent extraction, crystalline silicotitante non-elutable ion exchange, and small tank tetraphenylborate precipitation.[2]

---

[2] A fourth option, direct disposal in grout, was previously considered by SRS. Westinghouse Savannah River Company eliminated direct grout because the schedule uncertainty due to public and regulatory approval and potential litigation did not meet a 2010 schedule requirement based on available tank farm space. Fortenberry, J. K. 1998 (November 20). Memorandum to G. W. Cunningham regarding the SRS Report for Week Ending November 20, 1998. http://www.dnfsb.gov.

SMALL TANK TETRAPHENYLBORATE PRECIPITATION

The small tank tetraphenylborate precipitation process uses a sodium tetraphenylborate (NaTPB) reagent to remove cesium from the high-level waste salt solutions. The processing approach is fairly straightforward: The NaTPB reagent is added to a batch of salt solution and stirred to promote the formation of a cesium tetraphenylborate (CsTPB) precipitate, which is subsequently separated from solution by filtration. The precipitate is washed to remove unreacted NaTPB and excess salt and is then sent to the Defense Waste Processing Facility (DWPF) for further processing[3] and immobilization in glass. The decontaminated salt solution is sent to the on-site saltstone facility.

The STTP option is an engineered version of the in-tank precipitation (ITP) process that was originally designed and demonstrated for cesium removal at Savannah River. The ITP process was designed to be carried out in an existing high-level waste tank at Savannah River, but the process was abandoned after a large benzene excursion was observed during the startup of processing operations in Tank 48 in 1995 (NRC, 2000, pp. 44-49).[4] The STTP process, as currently designed, will be carried out in specially designed tanks that are smaller than the existing tanks at Savannah River to reduce contact time between the salt solutions and the NaTPB reagent, which should reduce the decomposition of tetraphenylborate and the generation of benzene and also to allow the safe handling and abatement of any benzene that is generated during processing.

The process, if implemented, will be carried out on 67 production batches of waste obtained by transferring supernate and dissolved salt cake from one or more million-gallon high-level waste tanks. SRS will obtain detailed compositional data on each waste batch and will run additional tests to confirm its compatibility with the selected processing option before the batch is processed.

2000 NRC Committee Recommendations

The 2000 NRC committee had several concerns regarding the

---

[3]The CsTPB precipitate is treated with acid, which allows for controlled release of benzene. The benzene is separated and burned at an on-site incinerator. The remaining aqueous stream contains boron, cesium, and potassium salts, which are suitable for vitrification in the DWPF.

[4]The benzene excursion was produced by decomposition of TPB, most likely through reactions with a metal catalyst in the tank waste. This excursion has been difficult to explain, and subsequent lab tests have not been able to reproduce it.

STTP option. First and most important, the committee determined that considerable effort was needed to identify the mechanism of tetraphenylborate decomposition, including the estimation of bounding TPB catalytic decomposition rates. In addition, a number of other issues were identified:

- whether the process can achieve the required decontamination factors (DFs)[5] of 7,700 (average) and 40,000 (upper bound) for Cs-137,
- whether the composition of the process stream to the vitrification process in the Defense Waste Processing Facility is acceptable for making glass,
- whether washing and recycle of the process stream can minimize the amount NaTPB required for this option,[6]
- whether foaming of the waste after treatment with NaTPB could block transfer lines or result in poor separation of CsTPB from solution, and
- whether cycle times and products associated with the decomposition of CsTPB during precipitate hydrolysis processing are compatible with existing processes.

The 2000 NRC committee recommended that as part of the effort to bound catalytic decomposition rates, SRS should develop robust testing protocols to process moderately sized samples of real waste from each of the blended batches from the high-level waste (HLW) tanks at SRS using NaTPB. The 2000 NRC committee also recommended that tests on moderately sized samples of real waste be implemented as soon as possible to help assess the viability of the STTP option.

Current Research and Development Results

SRS appears to be making significant progress to resolve the issues raised in the 2000 NRC committee report, and the present committee commends the STTP team for its research accomplishments. In particular, SRS is making good progress to (1) further elucidate the general features of the catalytic decomposition of sodium tetraphenylborate, using various analytical techniques (nuclear magnetic resonance [NMR], extended x-ray

---

[5]Decontamination factor (DF) is the ratio of feed solution contaminant concentration (in this case cesium-137) to the contaminant concentration of the solution after treatment by the cesium removal process.
[6]Design of the small tank option has focused on minimizing the amount of NaTPB reagent used in the cesium separation process. Thus, work has concentrated on the establishment of conditions that optimize the recovery of excess NaTPB during washing and the recycling of this reagent.

absorption fine structure [EXAFS], and transmission electron microscopy [TEM]) to help define the nature of a Pd/Hg catalyst system and the TPB decomposition intermediates, and (2) design, construct, and demonstrate equipment for preliminary testing of real waste in 0.5-, 2.0-, and 20-liter samples. The committee offers comments on this work below.

Mechanism of TPB Decomposition

In 1997, SRS personnel uncovered the possible role of metal catalysts in TPB decomposition and subsequent release of benzene. SRS personnel have indicated that Pd/Al_2O_3 in the presence of mercury (especially diphenylmercury [$HgPh_2$]) can function as a catalyst system for TPB decomposition in both simulants[7] and in real waste samples. Consultants hired by SRS have provided a plausible, but speculative, mechanism for Pd/Hg-catalyzed TPB decomposition that shares some of the features of catalytic Suzuki coupling (Miyaura and Suzuki, 1995).

Analysis by ^{11}B NMR spectroscopy has revealed some additional features of the catalytic decomposition of TPB by Pd(0) on alumina/$HgPh_2$, including the observed induction period.[8] Mercury may participate in catalyst activation by facilitating the nucleation and growth of palladium or Pd/Hg nanoclusters. The reactivity of nanoclusters is a function of surface area. The presence of mercury may enhance the reactivity of paladium particles as a catalyst for TPB hydrolysis. EXAFS and TEM analyses completed on simulant solids reveal face-centered-cubic (*fcc*) palladium nanoclusters and palladium-rich *fcc* clusters with mercury atoms surrounded by palladium atoms; both nanoclusters appear to be stabilized against aggregation by other components in HLW simulants.

Although a catalyst system for TPB has been identified, observed decomposition rates, using the best estimates of SRS tank waste concentrations of palladium and mercury, are still too low to account for the 1995 Tank 48 excursion. The average rate for the Tank 48 excursion, estimated at 10 mg of benzene per liter-hour of salt solution, was achieved using 26 mg/L of the catalyst Pd(0) on alumina in combination with Hg(II). It should be noted, however, that palladium concentrations in the tanks are currently estimated to range from 0.01 to 1.5 mg/L. The incubation periods have not been reproduced consistently with simulants or real waste, and the synergetic effects of mercury or diphenylborate are

---

[7]Simulants are laboratory-developed materials that mimic the properties of real waste samples. They are similar to real waste samples except that they do not contain radioactive species.

[8]The Tank 48 benzene excursion occurred 3 months after NaTPB was added to the waste. This incubation period is thought to be important for catalyzing TPB to produce benzene.

not yet clear. Moreover, it is puzzling that no significant TPB decomposition was observed in any of the tests conducted with real waste by SRS, even after 6 months, except for one run conducted with an added catalyst simulant. Thus, although progress has been made, there is still an insufficient understanding of the TPB decomposition process to allow rates to be bounded with a high level of confidence. This is especially true given the compositional variability of the tank waste to be processed.[9]

> **Finding: The process by which TPB decomposes is not completely understood and *is not predictable*, either mechanistically or empirically. At least two issues remain: (1) bounding rates for the generation of benzene, and (2) possible DF loss due to unanticipated rapid TPB decomposition.**

The committee believes that both of these issues can be engineered around,[10] but that cost and schedules may be impacted. Specifically, the lack of understanding and predictability of the TPB decomposition process presents the possibility that one or more batches of waste may not be treatable without unplanned-for tank waste blending to dilute catalysts present in the waste.[11] A commitment by SRS to put into place preprocessing testing to confirm the suitability of each batch of feed would significantly reduce the potential impacts and effects of untreatable batches, although it would leave unresolved the issue of how such batches should be processed. SRS's plans to isolate roughly million-gallon batches of real waste for testing prior to processing by these types of procedures partially circumvent the need for a full understanding of the reasons behind the 1995 excursion in Tank 48.

> **Recommendation: Given the lingering uncertainties in the catalytic decomposition of TPB, the committee believes that additional research should be undertaken to improve the predictability of TPB decomposition if STTP survives downselection.**

---

[9] A discussion of SRS tank waste compositions is provided in NRC (2000). Concentrations of many key radionuclides (e.g., cesium and actinides) in the tanks vary by several orders of magnitude, and concentrations of trace constituents that might act as catalysts are poorly known.

[10] As discussed in NRC (2000), the use of specially designed processing tanks will allow for much better control of reaction rates and benzene handling than is possible using the existing underground tanks as was planned for the ITP process.

[11] Given the variability in tank waste compositions, it may be possible to blend an unacceptable batch with waste from another tank to dilute the catalyst concentrations.

Waste Foaming

Waste foaming was first observed in laboratory tests on real waste samples and is caused by the entrainment of air during stirring of NaTPB-waste slurries (see NRC, 2000, p. 49). Foaming is of concern because of the potential for clogging process transfer lines and inhibiting the separation of CsTPB from the waste slurry. Antifoaming and defoaming agents[12] have been identified and tested using HLW simulants. Although an effective antifoaming agent (IITB52)[13] has been identified, its stability is limited, its delivery system has not been established, and the downstream consequences of its use for processing have not yet been evaluated.

A number of antifoam issues remain. SRS plans to evaluate the effects of radiation on the performance of IITB52 by conducting a series laboratory experiments using irradiated and unirradiated samples. Additionally, SRS is planning to conduct process simulation studies on the IITB52 antifoam agent to determine its effects on downstream processes. This agent will also be utilized in a test on real waste.

> **Recommendation: Further research on antifoaming agents, including diluents,[14] agent stability under processing conditions, and their impact on downstream processing is recommended. Determinations of the extent to which different batches of the antifoaming agent will perform under operating conditions with real waste also should to be made.**

This testing will allow process performance to be established systematically and evaluated under conditions that safely bracket the acceptable conditions for planned processing operations.

Real Waste Tests

Tests of the STTP process of approximately 6 months' duration using real waste have been developed and completed on samples from six different tanks. Tests conducted without added Pd(0) on alumina/$HgPh_2$ showed little TPB decomposition and achieved sufficient cesium DFs (>40,000). However, these tests were conducted on samples of waste

---

[12]These agents are chemical compounds that reduce the viscosity of the waste slurry, thereby inhibiting the development of air bubbles.
[13]IITB52 is a water-soluble liquid mixture of esters with a density of 1.01 g/mL.
[14]The diluent used for IITB52 is wash water.

supernate only and did not include dissolved salt cake.

Twenty-liter continuous stir tank reactor[15] demonstrations with real waste have been run successfully, and rapid cesium and strontium decontaminations have been demonstrated.[16] In one test with added catalyst simulant (7.8 mg/L $Pd(0)/Al_2O_3$ + 85 mg/L Hg(II)), acceptable cesium, strontium, and uranium DFs were achieved and maintained.[17] Benzene monitoring and abatement have been demonstrated.

> **Finding: Based on the real waste tests, the STTP process appears to meet cesium DF requirements for SRS tank waste.**
>
> **Recommendation: SRS should continue to refine preprocessing testing protocols, and if STTP is selected as the primary or backup option, plans should be made to process moderately sized samples from *each* of the proposed processing batches using MST and TPB and the selected antifoaming agent before the process is implemented.**

This preprocessing is recommended to mitigate the possibility of unplanned-for tank waste blending to dilute TPB decomposition catalysts present in the waste.

Committee Conclusions on the STTP Process

The STTP process has remaining technical uncertainties, but engineering solutions can probably be found for most of these potential processing problems. However, because of the unpredictability of the decomposition rate of TPB, there remains the risk that one or more of the 67 HLW production batches will require additional or special treatment before it can be processed using this option.

---

[15] A continuous stir tank reactor maintains the same concentration of reactants throughout the tank by stirring.
[16] DF_{Cs} >40,000; DF_{Sr} ~100.
[17] The DF for cesium was maintained between 10,000 and 40,000.

CRYSTALLINE SILICOTITANATE ION EXCHANGE

The ion exchange process for removing ionic species from aqueous solutions has been in commercial use for more than 100 years. Although the underlying technology is well established, ion exchange for cesium removal from high-level waste presents many challenges. The ion exchange material must withstand both high alkalinity and high radiation fields while exhibiting selectivity for cesium in the presence of much higher concentrations of sodium and potassium. A promising ion exchange material, crystalline silicotitanate (CST), has been the subject of extensive R&D efforts at numerous laboratories over the last 30 years and is the material of choice at SRS for separation of cesium from high-level waste by ion exchange.

In the current design for this process, CST will be packed into three 5-foot-diameter by 16-foot-long columns arranged in series. The columns will be cooled by the flow of process liquids through the column, which will remove heat produced by radioactive decay. The salt solutions will be filtered and then pumped through the first column, known as the lead column, at moderate pressure, where most of the cesium is expected to be exchanged. The solution will exit that column into a gas separation apparatus, where radiolytic gas (mainly hydrogen) will be removed. The solutions will then be pumped through the second and third columns, known as the middle and guard columns, for further decontamination if necessary, and the decontaminated solutions will be sent to the saltstone facility for immobilization in grout.

CST is nonelutable, so once the lead column is loaded with cesium it will be removed and sent to the DWPF for processing. The loaded CST will be removed from the column, size-reduced by grinding, mixed with glass frit, sampled, and transferred to the glass vitrifier. Once the lead column is removed from the processing facility, the middle column becomes the new lead column while the guard column becomes the new middle column. A column loaded with fresh CST is installed as the guard column. The facility is designed with valves and jumpers so that column positions can be swapped with a minimum of handling.

2000 NRC Committee Recommendations

Concerns by the 2000 NRC committee regarding CST as a process for cesium removal centered on material consistency and column design. The committee noted that CST displayed wide variability in performance, which could probably be traced to manufacturing variability, variations in pretreatment[18] of the material at SRS, or in the testing used to characterize the materials. As a result, the 2000 NRC committee recommended that efforts be made to ensure that a consistent and reproducible material could be obtained for use, and that uniform CST pretreatment and testing protocols be developed.

The 2000 NRC committee further concluded that the column design for the CST process may not be adequate for the thermal loadings and radiation fields expected with real waste. In addition, the committee determined that possible problems with radiolytic gas generation inside the ion exchange column, which could disrupt the flow of liquid and reduce the efficiency of the ion exchange process, had not been resolved. As a result, the 2000 NRC committee recommended that the ion exchange column design be reevaluated. In addition, the committee recommended that an R&D effort be undertaken to study factors that are important for process and column operation and design.

Current Research and Development Results

The R&D program in place for CST is designed to address issues of concern identified by the 2000 NRC committee. Research thrusts related to the CST process include the following:

- the chemical and thermal stability of manufactured CST, including chemical pretreatment requirements;
- the effects of gas generation in CST ion exchange columns; separation of radiolytic gas from the liquid process streams during transfers between columns; and handling, size reduction, and sampling of CST in the preparation of the DWPF feed;
- the performance of CST in columns, the kinetics of sorption related to temperature and feed composition, the capacity of CST to load cesium, modeling of CST performance, and alternatives to conventional column designs and parameters; and
- the impact of additional titanium from CST on the DWPF glass product.[19]

---

[18]The form of CST for use with tank wastes is described in the literature as a sodium salt, but it is manufactured and distributed in a protonated form at pH 3; pretreatment is needed to convert it to the sodium salt. The 2000 NRC committee concluded that this pretreatment may contribute to the observed instability of CST.

[19]Because CST is nonelutable, it must be incorporated into the high-level waste glass made at the DWPF. For further discussion of the possible impact of additional titanate from CST on DWPF glass, see the discussion on additional titanium later in this section on CST.

CST Performance and CST Pretreatment Technologies

The tank wastes at SRS are caustic and contain significant amounts of aluminum. Previous studies showed that when CST, which is delivered to SRS in an acid form, is contacted with caustic waste solutions, the rise of pH leads to $Al(OH)_3$ precipitation and immediate plugging of the ion exchange columns. Preconditioning of the CST columns with aluminum-free NaOH solutions appeared to be an easy answer to the problem. This work showed, however, that column plugging can be caused by precipitates other than aluminum hydroxide.

Testing at ORNL and SRS has shown that niobium introduced into CST during manufacturing could be dissolved and reprecipitated as several niobium phases (sodium-niobiate, sodium-sulfate-halide, cancrinite) at elevated temperature and thereby plug the ion exchange column. As a consequence, manufacturing processes for CST, including preconditioning to remove leachable niobium, have been modified. These modifications resulted in removal of much (>95 percent) of the leachable niobium added during manufacture and the removal of about 40 percent of the leachable silica. The low residual niobium eliminated precipitation of hydrous niobium oxides as a cause of column plugging.

Silica leaching from CST also has been found to impact column performance through the formation of an aluminosilicate precipitate, even after leachable silica is removed from the CST as described previously. These precipitates include cancrinite and minor amounts of sodalite. In addition, the precipitation of aluminosilicates on CST was observed in waste simulant experiments (described below) at elevated temperatures (50-120 °C) and/or on prolonged exposure (86 days) to the simulants.

Although fundamental understanding of the formation and impact of secondary minerals and phases has been enhanced through research, particularly with respect to the formation of new niobium phases, significant questions remain regarding aluminosilicate precipitation. The absence of useful predictive models and basic understanding of the events leading to precipitation of aluminosilicates has inhibited work on alternative designs for the CST process. This is discussed in more detail below.

Finding: SRS does not appear to have a clear and comprehensive understanding of the mechanism of aluminosilicate precipitation on CST. This issue poses potentially high technical risk for this candidate process.

CST is manufactured by only one company in the world, UOP. This supplier has worked cooperatively with SRS to improve the manufacturing process to remove the niobium-based impurity phase and decrease the amount of leachable silica. Nevertheless, if this supplier were to cease manufacturing of CST, it might prove difficult for SRS to identify an alternate source of this material. In addition, it is unlikely that a new supplier would be familiar with the manufacturing process developed by the present supplier, parts of which are proprietary. As a result, it could take time to develop new and reproducible manufacturing and testing protocols, which could result in substantial delays in the program.

Finding: The reliance on a single supplier for CST poses potentially high schedule risks for this candidate process.

Chemical and Thermal Stability of Cesium-Loaded CST

R&D by SRS on the characteristics of cesium-loaded CST has the following two objectives: (1) to understand the principles and resolve issues related to temperature changes during processing in order to determine the time and temperature profile at which irreversible desorption of cesium from CST occurs after CST is added to waste simulants,[20] and (2) to determine the reason for the apparent reduced cesium capacity of CST following long-term exposure to cesium-bearing simulated waste solutions. In these experiments, cesium contained in a nonradioactive waste simulant is loaded onto CST at room temperature.

CST shows a slight loss of capacity after exposure to cesium solutions for more than 9 months. Loading of CST with cesium from simulated waste was found to be proportional to carbonate concentration, but the reason for this effect is not understood. Irreversible sorption of cesium at temperatures greater than 50 °C remains an unexplained phenomena, and factors that determine the reversibility and irreversibility of cesium sorption are not yet clearly understood. SRS representatives told the committee that this problem can be eliminated by controlling the operating temperature of the columns so that it does not exceed 50 °C.

Finding: Thermal stability most likely will not pose an insurmountable problem if operating temperatures remain low and CST columns are changed before they lose their sorptive capacity. However, the operating margins are small.

---

[20] As discussed in NRC (2000), at temperatures above 50 °C, cesium has been observed to undergo irreversible deposition from CST. The reasons for this desorption are unknown.

The effect of organic impurities and minor components on cesium sorption on CST was also examined by SRS to determine the effect of organic compounds, carbonate, oxalate, and peroxide on the rate of cesium loading on CST. Organic impurities, and oxalate, did not affect column performance. Cesium loading on CST was found to be proportional to carbonate content, as noted previously. High concentrations of peroxide were observed to decompose CST, although SRS does not expect this to be a problem at anticipated operating conditions at SRS (see below).

Finding: Significant progress has been made in understanding the effects of chemical impurities on the performance of CST.

CST Process Columns

R&D has been focused on determining whether gas generated by water radiolysis within a CST ion exchange column can affect cesium sorption. In addition, the performance of the CST column, including the performance of alternative column configurations, has also been examined. Measurements of cesium sorption in the presence and absence of radiolytic gas showed no significant differences. Radiolysis will generate hydrogen in concentrations that are expected to exceed the explosive limit. SRS notes that the process columns are expected to be vented and that gas generation is a manageable safety matter.

Tests of the effects of gas generation in the tall columns that are part of the current conceptual design were conducted using hydrogen peroxide.[21] These studies showed that hydrogen peroxide reacts with CST to liberate silica and titanium and form aluminosilicate precipitates on the surfaces of the CST particles. The formation of these precipitates plugged the columns and hindered hydraulic removal of the CST. SRS calculated that the amount of hydrogen peroxide formed by radiation in a loaded process column would be much lower than the amount used for these tests and, therefore, precipitate formation and column plugging from this source would not represent a significant risk during actual processing operations.

SRS has also demonstrated that disengagement of gas during transfers of liquids between ion exchange columns would be adequate to reduce the transfer of gas from the lead to the middle column and therefore would represent a low risk. In light of the severe operational consequences attending interactions of CST and hydrogen peroxide—the plugging of an ion exchange column could lead to the loss of fluid flow

---

[21]Hydrogen peroxide will be formed by water radiolysis in a column loaded with cesium.

and the buildup of heat—these reactions could be a high risk during actual processing operations.

SRS is also evaluating alternative methods of contacting CST with the process solution. Current column designs are based in part on CST transfer requirements to the DWPF. Other designs, such as shorter columns or moving beds, may be difficult to implement reliably.

> **Finding: The present conceptual design for this process requires large sorption columns in series that are potentially subject to blockage, precipitate formation, and gas bubble formation. These phenomena may disrupt flow and sorption of cesium in the columns.**

As noted above, these phenomena have been studied individually and some of the problems have been solved. However, the committee has not seen evidence that SRS is attempting to integrate the solutions to these various problems into a relevant process simulation.

Effect of Additional Titanium from CST on the DWPF

The tolerance of DWPF glass to the addition of titanium from CST and also from monosodium titanate (MST) expected to be used for removal of strontium and alpha emitters (discussed below) has been reexamined by SRS. The presence of titanium can lead to crystallization in glass, which in turn can increase the liquidus temperature.[22] Crystallization in the DWPF melter can lead to processing problems.

Glasses were cooled at different rates to study their product consistency (Edwards, 2001). These glasses contained CST and MST (plus a simulated sludge) in amounts consistent with operations where real waste would be treated by both CST and MST. The measured values of these glasses consistently fall above the predicted values in the models used to predict durability. On the basis of these test results, SRS concluded that the very good durability of the CST-containing glasses implies that durability may not be the limiting factor for waste loading in the CST option for cesium removal, or the MST option for actinide or strontium removal.

> **Finding: The presence of titanium in presently estimated amounts does not appear to negatively impact the quality of the DWPF glass product.**

---

[22]The temperature at which there is an equilibrium between the glass and the primary crystalline phase.

If the final flowsheet for the CST or the MST process adds more titanium than currently envisioned, then the durability of the glass may have to be reevaluated.

Discussion

The need for information to allow process design and application to proceed has not been met adequately, as noted in the preceding discussion. Although alternative column designs and gas evolution issues have generally been identified, the committee did not learn of design selections or specifics of the management of safety issues associated with radiolytic gas formation and management. Suggested remedies for the plugging of columns observed in small-scale experiments have not yet been tested on a larger scale or with real waste. Most of the experimentation with waste simulants failed to include trace elements likely to be found in real waste.

> **Finding: Information needed to evaluate the risk of applying CST to cesium removal from the treated supernate has not been fully developed, although many of the issues have been identified. Therefore, the technical uncertainties remaining for the application of CST—including column plugging, resistance to hydraulic transfer, irreversible desorption, and column system technologies—will constitute a high risk for the use of this process for cesium removal.**
>
> **Recommendation: If CST is selected as either the primary or the backup option, the technical uncertainties identified above must be addressed, particularly alternative column designs to mitigate aluminosilicate buildup and radiolytic gas formation.**

Committee Conclusions on the CST Process

Of the three cesium separation processes under consideration, it is the committee's judgment that CST has the most technical uncertainties and the highest technical risks.

CAUSTIC SIDE SOLVENT EXTRACTION

As a general method of separation, solvent extraction is a mature technology that has been used in the nuclear industry for more than 50 years, although primarily with acidic process streams. At SRS, the goal of this process is to extract cesium ions from the aqueous waste stream into a immiscible solvent, thereby reducing the radionuclide content of the aqueous phase and enabling it to be disposed in the saltstone facility. The caustic process under development at SRS would be the first plant-sized caustic solvent extraction unit at Savannah River.

The basic principle of CSSX is to use a sparingly soluble diluent material that carries an extractant that will complex with cesium ions in the caustic solution. Separated cesium can then be stripped back into an aqueous phase ready for transfer to DWPF. Following cesium extraction, the solvent is scrubbed with dilute caustic to remove other salts from the solvent stream. The solvent is then contacted in a countercurrent flow with a dilute acid stream to transfer cesium to the acid stream (in the strip stages). The solvent is then scrubbed or purged to remove degradation products prior to recycling to the front of the process.

2000 NRC Committee Recommendations

The 2000 NRC committee found that although solvent extraction in general is a well-developed process, the technical maturity of the proposed solvent extraction process for the removal of cesium from high-level waste at SRS lagged significantly behind that for the two competing processes (STTP and CST).[23] As a result, the majority of the 2000 NRC committee's conclusions and recommendations focused on operational concerns.

These operational concerns were numerous. The committee questioned whether the solvent system could be cleaned and reused successfully. The buildup of particulate matter at the interface of the two solvent phases, common in solvent extraction, was considered by the previous committee to be a possibility in the process at SRS and could limit separation of these phases. The impact of changes in the feed

---

[23]The solvent used in cesium extraction studies at SRS is a multicomponent system. BoBCalixC6 is a calixarene crown ether that appears to work through a combination of effects to generate a cavity that preferably incorporates cesium relative to other ions. BoBCalixC6 is present in the CSSX solvent system in 0.01 M concentration. A diluent modifier, Cs-7SB, increases cesium extraction and is present at 0.5 M. An inhibitor, tri-N-octylamine, that inhibits impurity effects, is present at 0.001M. Isopar L is the solvent for these three active organic constituents of the extractant.

composition, pH, temperature, and ratios of the solvent constituents on the separation efficiency and capacity of the solvent system was also of concern. Other areas of concern were whether sophisticated control systems would be required to maintain a steady-state, high-DF operation, and the effect of ions in tank waste on separation efficiencies. The 2000 NRC committee also questioned whether a reliable supply (and supplier) of the calixarene crown ether (BobCalixC6) used in the CSSX process at SRS would become available.

As a result of these concerns, the 2000 NRC committee had three recommendations regarding the use of this process. First, a "cold" demonstration of this process on a modest scale was recommended. The purpose of this demonstration was to address as many of the aforementioned operational concerns as possible and, in particular, to identify any possible "showstoppers" that would preclude the use of this process at SRS. Second, the committee recommended that design of a hot laboratory demonstration process, using real tank waste, on a scale sufficient to define the final process should begin as soon as the cold tests demonstrated a high degree of confidence in the feasibility of the process. Finally, the committee recommended that work begin immediately on defining the production capability and economics for commercial quantities of the calixarene crown ether.

Current Research and Development Results

SRS appears to have implemented a robust R&D program that addresses many of the operational concerns expressed by the 2000 NRC committee. A demonstration of the complete CSSX process flowsheet using simulated waste was conducted, and real waste tests are scheduled for the spring of 2001. The R&D program is addressing factors such as the chemical and physical properties of the solvent, the stability of the solvent system, batch DF extraction and stripping, and solvent commercialization. Details are provided below.

CSSX Proof of Concept

The overall CSSX process, including solvent recycle, has been demonstrated using simulants on the bench scale. SRS personnel reported that under these conditions, the desired goal of a DF of greater than or equal to 40,000 was met, as was a cesium concentration factor of about

15.[24] The R&D program has examined effects of impurities and trace components and effective temperature control, and SRS personnel report that no significant problems were encountered.

The operation of the centrifugal contactors (designed to provide efficient mixing between the solvents followed by separation of the phases) was also studied. SRS personnel determined that due to a problem with stage efficiency, a better understanding of the multistage hydraulic performance of the 2-cm centrifugal contactor used in the bench-scale tests was needed, but because this issue is mainly a result of the small size of the contactor, it will not be an issue in plant scale operations.[25]

Solvent Chemical Stability

Research on the chemical stability of the CSSX solvent system examined the solvent and how it works, phase behavior of the primary solvent components, and distribution performance[26] and solvent cleanup. The overall process was reported by SRS personnel to be very effective. The CSSX solvent system was found to be stable to precipitation of solids for at least a year. The lower limit of the formation of dense liquid organic phases, a detriment common in solvent extraction processes, was found to be at 20 °C, well below the planned operating temperature for this process (about 35 °C). The distribution of cesium between phases was found to be reproducible for the various steps in the CSSX process.

Research on the effects of impurities on the solvent also has been conducted. The solvent has been cycled through multiple extraction, scrubbing, and stripping batch processes using a waste simulant to determine whether impurity buildup would degrade stripping performance. An impurity buildup was observed to occur, but a dilute sodium hydroxide wash was found to be effective for cleaning the solvent system and removing these impurities. The waste simulants included noble metals and organic compounds that are expected to be encountered in real waste. These were not observed to degrade the solvent system or its performance.

Thermal stability tests were also undertaken to examine the operational limits of the solvent system. For each step of the CSSX process (extraction, scrubbing, and stripping), the stability of the solvent in contact with the waste simulant was analyzed with external heating to 35-60 °C. SRS determined that even at 110 days at the planned maximum

---

[24]The cesium concentration factor is the cesium concentration in the aqueous strip effluent [EW] divided by the cesium concentration in the feed simulant [FS]: [EW]/[FS].
[25]Plant-size contactors will be much larger and, as a result, will not have these hydraulic problems.
[26]Distribution performance measures the cesium distribution ratio between the organic and aqueous phases over the progressive steps in the CSSX process (extraction, scrubbing, and stripping).

operating temperature of 35 °C, performance remains good. SRS personnel also reported that the solvent does not undergo unacceptable degradation due to nonradiation effects such as the presence of noble metals.

Solvent Radiological Stability

The first purpose of solvent irradiation experiments is to determine the radiolytic stability of the solvent, identify decomposition products, and assess their impact on solvent performance. A secondary focus of these experiments is to compare simulant and real waste test data and to compare internal and external irradiation results. The radiolysis program involves four experiments: (1) cobalt-60 external irradiation of the solvent; (2) solvent self-irradiation with ^{137}Cs-spiked simulant; (3) contactor hydraulic performance with ^{137}Cs-spiked simulant; and (4) solvent self-irradiation with real waste from SRS.

Results of the radiolytic stability tests indicated no significant degradation of the solvent system. External irradiation of the solvent using cobalt-60 produced only minor decomposition.[27] Irradiation of the solvent using a ^{137}Cs-spiked simulant has not identified any radiolytic concerns.[28] In both sets of experiments, third-phase formation or formation of particulates at the solvent interface was not observed, and cesium decontamination efficiency was still found to be within the expected range.

Results from the contactor hydraulic performance test have not identified any radiolytic concerns. Physical observations did not identify any dose-related impacts, and chemical analysis of the samples from this test were in progress at the time of the committee's briefings.

Further tests of the solvent are planned in the spring of 2001 using real waste. Batch extraction tests, designed to measure the solvent performance using real waste from the SRS tank farms, will attempt to operate the process for 28 solvent turnovers, and demonstrate a decontamination factor of >15,000.

> **Finding: R&D on the CSSX process has, so far, encountered no significant technical obstacles, and there do not appear to be any technical obstacles to**

---

[27]Approximately 10 percent of BOBCalixC6 was lost at a 16-Mrad dose (equivalent to exposure of the solvent system to SRS real waste for 160 years), and about 10 percent of TOA was lost at a 6-Mrad dose (equivalent to exposure of the solvent system to SRS real waste for 60 years).

[28]In these experiments, the solvent system received exposure to cesium-137 equivalent to a decade of plant operation of the process with SRS real waste.

> **scale-up of the CSSX process to plant-scale operations.**

The research performed over the last year has addressed most of the issues identified by the 2000 NRC committee and appears to confirm the viability of this process at the laboratory scale. The present committee has been impressed with the unimpaired performance of this process under very rigorous testing. The strong success of the CSSX process in this R&D program is unusual for most process development.

> **Finding: Successful bench-scale demonstration of the complete process with actual tank waste is critical for qualifying the CSSX process for serious consideration in the down-selection process.**

This demonstration, if done well, will show whether the CSSX process can remove cesium from real waste at levels sufficient for saltstone requirements and whether pilot-scale testing is warranted.

Solvent Preparation and Commercialization

A key technology issue that impacts successful use of the CSSX process is the availability of the solvent at plant-scale quantities. Commercial suppliers of all solvent components, including BOBCalixC6, have been identified by SRS. BOBCalixC6 with greater than 97 percent purity has been prepared successfully by an outside manufacturer, and its production scale and cost are well within the economic and schedule parameters (given the projected loss rates) for the use of this process at SRS.[29] Additional suppliers of BOBCalixC6 have been identified by SRS, and a U.S. patent protects the government's rights to grant a license to manufacture this material. No other component of the solvent system appears to present an economic or scheduling burden to the use of this process.

> **Finding: No significant economic obstacles for scale-up of the CSSX process appear to have been encountered so far.**

> **Recommendation: If this process remains a viable candidate for cesium removal, monitoring of the cost and potential suppliers of the reagents for this process should continue.**

---

[29]For a plant-scale charge of 4,000 liters of solvent, 46 kg of BOBCalixC6 would be required. The cost for this amount of BOBCalixC6 should be approximately $5 million, well within the $8 million budgeted for this reagent.

Discussion

After being far behind in level of development compared to the other two processes a year ago, there has been a major acceleration of work on the CSSX process, and the development gap between it and the others has been reduced markedly. The committee is impressed by the overall quality of the science on this alternative. The seamless integration of research from several laboratories is especially impressive.

Finding: The main source of concern regarding the viability of the CSSX process continues to be the stability of the solvent system. The process itself should be relatively straightforward to scale up, at least from a mechanical standpoint, because the process hardware has been proven in nuclear applications and there are no solid-handling steps.

The potential concerns include chemical and radiolytic stability, the possible detrimental effects of its breakdown products on the DF and the cesium concentration factor, and the possibility of performance-degrading effects of trace components in real waste that have not been included in the simulated feed tests.

Recommendation: Extensive testing of the most performance-critical components of the solvent (e.g., composition, pH, and temperature ranges the solvent would most likely encounter) should continue in parallel with the bench-scale process test using real plant waste in order to give the greatest possible assurance that the required separation performance can be achieved and maintained with any waste composition likely to be encountered. Successful completion of this program will allow concerns about the solvent system to be characterized as low risk.

Committee Conclusions on the CSSX Process

Unless tests with actual waste encounter new problems, the CSSX option for cesium separation presents, at present, the fewest technical uncertainties of any of the three cesium separation alternatives.

ACTINIDE AND STRONTIUM REMOVAL

The removal of strontium and actinides (especially plutonium and neptunium) is an important step in the high-level waste processing flowsheet at SRS. As presently envisaged, strontium and actinides will be removed from the salt solutions in all three of the cesium processing options discussed in this report. At present, the use of monosodium titanate is the method of choice. Although the mechanisms for strontium and actinide removal by MST are not well understood, it is presumed that an ion exchange reaction of the sodium ions in the MST takes place, primarily with cations in higher oxidation states (e.g., strontium, plutonium, neptunium, and uranium) but also, to a lesser extent, with monovalent cesium and potassium cations.

The three cesium removal options are designed to process waste streams that have been treated to remove actinides and strontium. SRS plans to remove these radionuclides at the "front end" of processing operations with CST and CSSX by batch contact of the waste solution with finely powdered MST.[30] Incoming salt solution from storage tanks containing entrained sludge solids is pretreated with MST to adsorb strontium and plutonium. The resulting slurry is filtered using a cross-flow filter, and the MST and sludge solids are to be sent to the DWPF for vitrification.

2000 NRC Committee Recommendations

The 2000 NRC committee found that two major issues had to be resolved before SRS could successfully implement MST for actinide and strontium removal: (1) whether strontium and actinide removal could be accomplished within saltstone limits and throughput rates required by the DWPF, and (2) whether the MST concentrations used to remove strontium and actinides would exceed compatibility limits for DWPF glass. The 2000 NRC committee recommended that R&D be performed to resolve these issues, that requirements reliable sources for the manufacture of MST be established, and that SRS look at alternatives to MST.

Current Research and Development Results

The 2000 NRC committee questioned the assumption that strontium-alpha separation is to precede cesium removal. This challenge provides the basis of some of the R&D activities described below.

---

[30] With STTP, the MST actinide and strontium removal process is carried out concurrently with cesium removal.

Actinide and Strontium Removal Requirements

Removal of alpha emitting elements and strontium is required to meet saltstone waste acceptance requirements, which can be defined in terms of the average and the highest bounding activities of the process streams.[31] Further, the throughput rates must meet downstream feed requirements, especially to keep the DWPF operational.

One important focus of the R&D program is elucidating actinide and strontium removal rates for MST. Simulated waste solutions at 5.6 M Na^+ containing known quantities of strontium, plutonium, uranium, and neptunium were used and, at controlled temperature, samples were drawn and analyzed for sorbate concentrations after removal of solids by filtration. The results are dependent on temperature and ionic strength. Based on the experiments with waste simulants, SRS reported the following:

- MST removal of strontium is adequate to meet saltstone requirements. The experimental DF reported for strontium was about 150, which exceeds the maximum required DF of 26,
- MST removal of neptunium is inadequate to meet saltstone requirements for waste in some of the tanks. The experimental DF reported for neptunium is 3.47, which is well below the maximum required DF of 33, and
- MST removal of plutonium is also inadequate to meet saltstone requirements for waste in some of the tanks. The experimental DF reported for Pu is 11.3, slightly below the average required DF for the tank waste of 12, and well below the maximum required DF of 55 for tanks with the highest plutonium concentrations.

SRS also reported that saltstone limits probably could be met by blending waste from different tanks to reduce the required DFs for neptunium and plutonium. To this end, SRS plans to prepare 67 separate batches for MST and cesium processing by blending together waste from different tanks. Careful blending will allow SRS to dilute the high radionuclide concentrations in the "problem" tanks, thereby reducing the DFs required to meet saltstone requirements. In fact, SRS personnel reported that by careful blending and MST processing, they can produce batches that meet saltsone requirements for neptunium and strontium, and

---

[31]The required DF (average/bounding) for plutonium/americium is 12/55, for uranium it is 1/1, for neptunium it is 1/33, and for strontium it is 5/26. Bounding values are decontamination factors for tanks with the highest concentration of the radionuclide in question.

that these same batches can be processed by MST to achieve required DFs for plutonium.

An issue related to the use of MST for actinide and strontium removal is the possibility that colloidal plutonium could exist in the tank waste. Such material would not be chemically removed by the MST, and colloidal particles would be too small to be removed by filtration. However, filtration experiments using varying membrane pore sizes of samples from several waste tanks did not indicate the presence of colloidal plutonium.

> **Finding: The blending of tank waste to produce 67 process batches and treatment by MST appears to meet the saltstone requirements for neptunium and strontium decontamination. Based on the information received by the committee, MST appears to be adequate to separate Pu, as long as there is no colloidal plutonium in the waste, but with little margin to meet saltstone requirements.**

A technical uncertainty that remains to be resolved is the kinetics of sorption on MST. This issue is particularly important because of the additional titanium present in MST (and also from the CST), when in the waste feed at the DWPF, may exceed acceptable limits on the amount of titanium in the glass waste form.

> **Finding: The maximum quantity of titanate allowable in the process stream to meet the titanium levels acceptable in the vitrified waste form remains a technical uncertainty.**[32]

Solid-Liquid Separation Studies

Once MST solids have been added to the salt solutions to sorb strontium and actinides, these solids (along with any sludge solids in the waste) must be separated from the liquids and transferred to the DWPF. Filtration is currently the baseline process for solids removal, and several studies have been performed to elucidate filtration performance. The objectives of these studies are (1) to confirm baseline sludge and MST cross-flow filtration performance[33] at pilot scale with simulated wastes,

---

[32] SRS has conducted durability tests on CST- and MST-loaded glasses.
[33] In cross-flow filtration, the process stream flow is tangential to the filter surface, thereby minimizing the buildup of solids on the surface that reduces filter efficiency.

and (2) to evaluate alternative solid-liquid separation technologies for their potential to reduce facility size.

The main objective of the pilot-scale filtration studies is to measure filtration rates for slurries containing simulated sludge and MST in large, prototypic equipment. The need for backpulsing[34] was evaluated and the extremes of the system were tested. A target cross-flow filter permeate flux of 0.22 gallons per minute per square foot is desired for concentrating slurries with up to 5 weight percent insoluble solids. The observed fluxes meet or exceed these design assumptions. Filtration experiments have been carried out at the pilot scale with highly promising preliminary results, and appear to indicate that filtration can be achieved within the requisite parameters required for all three candidate cesium removal technologies.

A focus of the R&D program has been to investigate several ways to increase MST and sludge filtration rates. Additives, both flocculants and antifoamants, have been found that improve filtration. Cross-flow filter tests with flocculants have shown a 1.3-fold improvement in filter flow rate over baseline. Alternative filtration technologies are under investigation but have not yet provided significant results.

Ongoing studies include tests using real waste samples, including cross-flow flux and rheology measurements and tests of flocculants with and without MST. Additional pilot-scale filtration testing will involve filtration tests using sludge only and MST only for two waste compositions, and a potential test using a flocculant. Planned experiments on alternative filtration techniques include settling and decanting tests, high shear filtration (centrifugal) tests, and centrifuge evaluation.

Alternatives to MST for Actinide and Strontium Removal

In November 1999, SRS personnel reported to the 2000 NRC committee that alternative processes for actinide and strontium removal were being investigated in case MST fails to meet expectations. One of the most promising alternatives is a precipitation method for strontium and actinide removal using sodium permanganate. In this process, removal of actinides occurs upon the precipitation of hydrated manganese oxide following the sequential addition of strontium nitrate, calcium nitrate, and sodium permanganate to the highly alkaline waste solutions. The likely mechanism for actinide removal involves adsorption, inclusion, and occlusion in the hydrous manganese oxide matrix. The sodium

---

[34]Backpulsing is a method for removing particles that have collected in the pores and on the surface of the filter membrane using a periodic reversal of the transmembrane pressure.

permanganate process achieved higher DFs for strontium than MST (199 versus 150), for plutonium (30.4 versus 11.3), for uranium (1.88 versus 1.14), and for neptunium (7.90 versus 3.47).

Two other alternative adsorbents to MST, sodium nonatitanate (ST) and SrTreat® (proprietary), were also compared to MST for their ability to remove strontium and actinides. Both SrTreat and ST exhibit rates of removal for strontium that are similar to MST (approximately 0.15 µg/L of strontium remain in solution after treatment versus approximately 5 µg/L for MST after 107 hours). For plutonium removal, ST has a similar rate to MST (approximately 3.5 µg/L of Pu remain versus 7 µg/L after 107 hours), while the rate for SrTreat® was significantly lower (approximately 90 µg/L of plutonium remain after 107 hours). Similarly, ST and MST had almost identical rates of neptunium removal (approximately 60 µg/L of neptunium remain after 107 hours), while the rate for SrTreat was again much lower (approximately 300 µg/L of neptunium remain after 107 hours). In general, ST showed a behavior parallel to that of MST. It was concluded by SRS personnel that MST kinetics are adequate to meet the baseline preconceptual design for each processing alternative and that the ST sorbent process and a manganese-based precipitation process provide promising backups for MST.

> **Finding: Two alternate precipitation processes are competitive with MST. These employ sodium nonatitanate, which behaves similarly to MST, and sodium permanganate.**
>
> **Recommendation: The backup processes, sodium nonatitanate and the sodium permanganate-based precipitation process, should be studied further. The R&D program for these two processes should be based on that developed for MST and should continue until MST processing can be demonstrated to meet the saltstone, DWPF throughput, and DWPF glass requirements.**

If one of these backup processes is found to be superior to MST, its substitution for MST will have to be done soon so as not to delay the implementation of the cesium removal processes. Resolution of the choice for this process is largely independent of the choice of a cesium separation process.

Committee Comments on the MST Process

All of the cesium separation processes depend upon a separate step to remove strontium, neptunium, and plutonium. Currently, that step uses MST. Because the success of this step is essential to all three of the processes for cesium separation, the committee believes that continued R&D on alternate processes for the removal of actinides and strontium is essential until MST processing can be demonstrated to meet the saltstone, DWPF throughput, and DWPF glass requirements.

REFERENCES

Edwards, T.B., J.R. Harbour, and R.J. Workman. 2001. Impact of Cooling Rate on the Durability on CST Glasses: A Nonproprietary Summary. Westinghouse Savannah River Company WSRC-TR-2001-00125, Revision 0, Aiken, SC. 12 pp.

Miyaura, N., and Suzuki, A. 1995. Palladium Crystallized Reactions of Organoboron Compounds. Chemical Reviews 95: 2457-2483.

National Research Council. 2000. Alternatives for High-Level Waste Salt Processing at the Savannah River Site, Washington, D.C.: National Academy Press.

Pacific Northwest National Laboratory. 2000. Savannah River Site Salt Processing Project Research and Development Program Plan, Revision 0. PNNL-13253. Richland, Washington.

APPENDIX A
Letters of Request for this Study

Department of Energy
Washington, DC 20585
April 13, 2000

Dr. Kevin D. Crowley
Director
Board on Radioactive Waste Management
National Research Council
2001 Wisconsin Avenue, N. W. Washington, DC 20007

Dear Dr. Crowley:

I would like to take this opportunity to thank you and your Committee Members for your extraordinary effort providing the Department with an independent technical review of alternatives for processing the high-level radioactive waste salt solutions at the Savannah River Site. We agree with your interim comments noting that additional research and development is required for each option, and we are proceeding with addressing your comments in our research and development plans for fiscal years 2000 and 2001. I am looking forward to receiving your final report this month so that we can make adjustments in the current plans if needed.

I believe that the complexity of the salt processing technology alternatives warrants your continued involvement in our continuing research and development
efforts. Therefore, I would like to request that you and your Committee continue to support the Department throughout the next year by providing us with your independent review of each technology road map, and the selection criteria.

Dr. Huntoon has tasked me, as the Deputy Assistant Secretary for the Office of Project Completion, to provide the leadership and program management for technology development and selection of a preferred treatment alternative. I am working closely with the Office of Science and Technology, as well as the DOE-Savannah River Operations Office, to make sure that this effort is adequately supported. An Action Plan has been prepared, and is enclosed, which provides details of the roles and responsibilities for the project. I will be providing the Assistant Secretary for Environmental Management with quarterly progress reviews on each of the technology activities throughout the ensuing months, and I propose that we follow those reviews with a briefing to your Committee to keep you abreast of the salt processing project's progress. Of course, additional briefings, meetings, and documentation will be made available to the Committee as you deem necessary to support your review.

Based on the current schedule, we would be seeking your Committee's review of the items identified above in early summer 2000. Due to the short time available between now and the anticipated time we require your support, may I suggest that you utilize your existing Committee to expedite matters.

Mr. Ken Lang of my staff will be contacting you directly to coordinate the details.
Mr. Lang can be reached at (301) 903- 7453.

Thank you in advance for your continued support of DOE. I look forward to
working with you in this endeavor.

Sincerely,

Mark W. Frei
Deputy Assistant Secretary
for Project Completion
Office of Project Completion

Enclosure

cc:
M. Gilbertson, EM-52
K. Picha, EM-22
G. Rudy, DOE-SR
K. Gerdes, EM-54
B. Spader, DOE-SR
J. Case, DOE-ID

Department of Energy
Washington, DC 20585
June 15, 2000

Dr. Kevin D. Crowley
Director
Board on Radioactive Waste Management
National Research Council
2001 Wisconsin Avenue, NW
Washington, D.C. 20007

Dear Dr. Crowley:

Thank you for your May 16, 2000, letter responding to my request that the National Research Council continue its support of the Department's high-level waste salt processing alternatives at the Savannah River Site.

I am pleased that you would like to continue to provide technical assistance to the Department throughout the planned research and development phase of this project, pending approval of the Board on Radioactive Waste Management and the National Research Council Governing Board.

Your proposals to (I) comment on the criteria that will be used to select a processing alternative; (2) evaluate the results of the research and development work that is undertaken on the candidate processing alternatives; and (3) provide the Department with an assessment of whether the technical uncertainties have been sufficiently resolved to proceed with downsizing the list of alternatives will meet our needs throughout the remaining research and development period.
I found the interim report you provided on your current evaluation to be particularly
useful in planning the research and development now underway, and I am confident that an interim report for this phase of the study will be valuable in the selection of alternative processing technologies.

Mr. Kenneth Lang of my staff is available to support you and the committee for this review. Mr. Lang can be reached at (301) 903-7453.

Thank you for your continued support of DOE.

Sincerely,

Mark W. Frei
Deputy Assistant Secretary
for Project Completion
Office of Environmental
Management

cc:
M. Gilbertson, EM-52
K. Picha, EM-22
G. Rudy, DOE-SR
K. Gerdes, EM-54
G. Boyd, EM-50
B. Spader, DOE-SR
J. Case, DOE-ID

APPENDIX B

Interim Report

EVALUATION OF CRITERIA FOR SELECTING A SALT PROCESSING ALTERNATIVE FOR HIGH-LEVEL WASTE AT THE SAVANNAH RIVER SITE: INTERIM REPORT

Committee on Radionuclide Separation Processes for
High-Level Waste at the Savannah River Site

Board on Radioactive Waste Management
Board on Chemical Sciences and Technology
Division on Earth and Life Studies
National Research Council

NATIONAL ACADEMY PRESS
Washington, D.C.

NATIONAL ACADEMY PRESS   2101 Constitution Avenue, N.W.   Washington, D.C. 20418

NOTICE: The project that is the subject of this report was approved by the Governing Board of the National Research Council, whose members are drawn from the councils of the National Academy of Sciences, the National Academy of Engineering, and the Institute of Medicine. The members of the committee responsible for the report were chosen for their special competences and with regard for appropriate balance.

This study was supported by Contract/Grant No. DEFC0199EW59049 between the National Academy of Sciences and the Department of Energy. Any opinions, findings, conclusions, or recommendations expressed in this publication are those of the author(s) and do not necessarily reflect the views of the organizations or agencies that provided support for the project.

Library of Congress Cataloging-in-Publication Data
or
International Standard Book Number 0-309-0XXXX-X
Library of Congress Catalog Card Number 97-XXXXX

Additional copies of this report are available from National Academy Press, 2101 Constitution Avenue, N.W., Lockbox 285, Washington, D.C. 20055; (800) 624-6242 or (202) 334-3313 (in the Washington metropolitan area); Internet, http://www.nap.edu

Printed in the United States of America
Copyright 2001 by the National Academy of Sciences. All rights reserved.

THE NATIONAL ACADEMIES

National Academy of Sciences
National Academy of Engineering
Institute of Medicine
National Research Council

The **National Academy of Sciences** is a private, nonprofit, self-perpetuating society of distinguished scholars engaged in scientific and engineering research, dedicated to the furtherance of science and technology and to their use for the general welfare. Upon the authority of the charter granted to it by the Congress in 1863, the Academy has a mandate that requires it to advise the federal government on scientific and technical matters. Dr. Bruce M. Alberts is president of the National Academy of Sciences.

The **National Academy of Engineering** was established in 1964, under the charter of the National Academy of Sciences, as a parallel organization of outstanding engineers. It is autonomous in its administration and in the selection of its members, sharing with the National Academy of Sciences the responsibility for advising the federal government. The National Academy of Engineering also sponsors engineering programs aimed at meeting national needs, encourages education and research, and recognizes the superior achievements of engineers. Dr. William A. Wulf is president of the National Academy of Engineering.

The **Institute of Medicine** was established in 1970 by the National Academy of Sciences to secure the services of eminent members of appropriate professions in the examination of policy matters pertaining to the health of the public. The Institute acts under the responsibility given to the National Academy of Sciences by its congressional charter to be an adviser to the federal government and, upon its own initiative, to identify issues of medical care, research, and education. Dr. Kenneth I. Shine is president of the Institute of Medicine.

The **National Research Council** was organized by the National Academy of Sciences in 1916 to associate the broad community of science and technology with the Academy's purposes of furthering knowledge and advising the federal government. Functioning in accordance with general policies determined by the Academy, the Council has become the principal operating agency of both the National Academy of Sciences and the National Academy of Engineering in providing services to the government, the public, and the scientific and engineering communities. The Council is administered jointly by both Academies and the Institute of Medicine. Dr. Bruce M. Alberts and Dr. William A. Wulf are chairman and vice chairman, respectively, of the National Research Council.

COMMITTEE ON RADIONULIDE SEPARATION PROCESSES FOR HIGH-LEVEL WASTE AT THE SAVANNAH RIVER SITE

MILTON LEVENSON, *Chair*, Bechtel International (retired), Menlo Park, California
GREGORY R. CHOPPIN, *Vice-Chair,* Florida State University, Tallahassee
JOHN E. BERCAW, California Institute of Technology, Pasadena
DARYLE H. BUSCH, University of Kansas, Lawrence
JAMES H. ESPENSON, Iowa State University, Ames
GEORGE E. KELLER II, Union Carbide Corporation (retired), South Charleston, West Virginia
THEODORE A. KOCH, E.I. du Pont de Nemours and Company (retired), Wilmington, Delaware
ALFRED P. SATTELBERGER, Los Alamos National Laboratory, Los Alamos, New Mexico
MARTIN J. STEINDLER, Argonne National Laboratory (retired), Downers Grove, Illinois

Staff
ROBERT S. ANDREWS, Senior Staff Officer, Board on Radioactive Waste Management
CHRISTOPHER K. MURPHY, Program Officer, Board on Chemical Sciences and Technology
LAURA LLANOS, Senior Project Assistant, Board on Radioactive Waste Management
TONI GREENLEAF, Administrative Associate, Board on Radioactive Waste Management

BOARD ON RADIOACTIVE WASTE MANAGEMENT

JOHN F. AHEARNE, *Chair*, Sigma Xi and Duke University, Research Triangle Park, North Carolina
CHARLES MCCOMBIE, *Vice-Chair*, Consultant, Gipf-Oberfrick, Switzerland
ROBERT M. BERNERO, Consultant, Bethesda, Maryland
ROBERT J. BUDNITZ, Future Resources Associates, Inc., Berkeley, California
GREGORY R. CHOPPIN, Florida State University, Tallahassee
RODNEY C. EWING, University of Michigan, Ann Arbor
JAMES H. JOHNSON, JR., Howard University, Washington, D.C.
ROGER E. KASPERSON, Clark University, Worcester, Massachusetts
NIKOLAY P. LAVEROV, Russian Academy of Sciences, Moscow
JANE C.S. LONG, Mackay School of Mines, University of Nevada, Reno
ALEXANDER MACLACHLAN, E.I. du Pont de Nemours & Company (retired), Wilmington, Delaware
WILLIAM A. MILLS, Oak Ridge Associated Universities (retired), Olney, Maryland
MARTIN J. STEINDLER, Argonne National Laboratories (retired), Argonne, Illinois
ATSUYUKI SUZUKI, University of Tokyo, Japan
JOHN J. TAYLOR, Electric Power Research Institute (retired), Palo Alto, California
VICTORIA J. TSCHINKEL, Landers and Parsons, Tallahassee, Florida

Staff

KEVIN D. CROWLEY, Director
ROBERT S. ANDREWS, Senior Staff Officer
BARBARA PASTINA, Staff Officer
GREGORY H. SYMMES, Senior Staff Officer
JOHN R. WILEY, Senior Staff Officer
SUSAN B. MOCKLER, Research Associate
TONI GREENLEAF, Administrative Associate
LATRICIA C. BAILEY, Senior Project Assistant
LAURA D. LLANOS, Senior Project Assistant
SUZANNE STACKHOUSE, Project Assistant
ANGELA R. TAYLOR, Senior Project Assistant
JAMES YATES, JR., Office Assistant

BOARD ON CHEMICAL SCIENCES AND TECHNOLOGY

KENNETH N. RAYMOND, *Co-Chair*, University of California, Berkeley
JOHN L. ANDERSON, *Co-Chair*, Carnegie Mellon University, Pittsburgh, Pennsylvania
JOSEPH M. DESIMONE, University of North Carolina and North Carolina State University, Raleigh
CATHERINE C. FENSELAU, University of Maryland, College Park
ALICE P. GAST, Stanford University, Stanford, California
RICHARD M. GROSS, Dow Chemical Company, Midland, Michigan
NANCY B. JACKSON, Sandia National Laboratory, Albuquerque, New Mexico
GEORGE E. KELLER II, Union Carbide Company (retired), South Charleston, West Virginia
SANGTAE KIM, Eli Lilly and Company, Indianapolis, Indiana
WILLIAM KLEMPERER, Harvard University, Cambridge, Massachusetts
THOMAS J. MEYER, Los Alamos National Laboratory, Los Alamos, New Mexico
PAUL J. REIDER, Merck Research Laboratories, Rahway, New Jersey
LYNN F. SCHNEEMEYER, Bell Laboratories, Murray Hill, New Jersey
MARTIN B. SHERWIN, ChemVen Group, Inc., Boca Raton, Florida
JEFFREY J. SIIROLA, Chemical Process Research Laboratory, Kingsport, Tennessee
CHRISTINE S. SLOANE, General Motors, Troy, Michigan
ARNOLD F. STANCELL, Georgia Institute of Technology, Atlanta
PETER J. STANG, University of Utah, Salt Lake City
JOHN C. TULLY, Yale University, New Haven, Connecticutt
CHI-HUEY WONG, Scripps Research Institute, La Jolla, California
STEVEN W. YATES, University of Kentucky, Lexington

Staff

DOUGLAS J. RABER, Director
RUTH MCDIARMID, Program Officer
CHRISTOPHER K. MURPHY, Program Officer
SYBIL A. PAIGE, Administrative Associate

Acknowledgement of Reviewers

This report has been reviewed in draft form by individuals chosen for their diverse perspectives and technical expertise, in accordance with procedures approved by the National Research Council (NRC) Report Review Committee. The purpose of this independent review is to provide candid and critical comments that will assist the institution in making the published report as sound as possible and to ensure that the report meets institutional standards for objectivity, evidence, and responsiveness to the study charge. The review comments and draft manuscript remain confidential to protect the integrity of the deliberative process. We wish to thank the following individuals for their participation in the review of this report:

Robert M. Bernero, U.S. Nuclear Regulatory Commission (retired)
J. Brent Hiskey, University of Arizona
Lawrence Kershner, Dow Chemical Company
James W. Mitchell, Bell Labs/Lucent Technologies
Kenneth N. Raymond, University of California, Berkeley
Edwin L. Zebroski, Elgis Consulting

Although the reviewers listed above have provided many constructive comments and suggestions, they were not asked to endorse the conclusions or recommendations nor did they see the final draft of the report before its release. The review of this report was overseen by Royce W. Murray, University of North Carolina, appointed by the NRC's Report Review Committee, who was responsible for making certain that an independent examination of this report was carried out in accordance with institutional procedures and that all review comments were carefully considered. Responsibility for the final content of this report rests entirely with the authoring committee and the institution.

Contents

SUMMARY	1
INTRODUCTION	1
BACKGROUND	2
DOE SELECTION CRITERIA AND GOALS	3
COMMENTS ON CRITERIA	6
FINDINGS AND RECOMMENDATIONS	7
REFERENCES CITED	8
APPENDIX A. BIOGRAPHICAL SKETCHES OF COMMITTEE MEMBERS	10
APPENDIX B. LETTERS OF REQUEST FOR THIS STUDY	13
APPENDIX C. HIGH-LEVEL WASTE TANKS AT THE SAVANNAH RIVER SITE	18

SUMMARY

At the request of the U.S. Department of Energy (DOE), the National Research Council formed a committee in 1999 to provide an independent technical review of alternatives selected by the Savannah River Site (SRS) for processing the high-level radioactive waste (HLW) salt solutions stored there. The final report of that committee, *Alternatives for High-Level Waste Salt Processing at the Savannah River Site*, was issued in August 2000. DOE subsequently asked the National Research Council to provide an assessment of DOE's efforts to select a processing alternative for removal of cesium, strontium, and actinides from high level waste at the Savannah River Site. A new committee was appointed, and it addresses in this interim report the first part of its statement of task— "evaluate the adequacy of the criteria that will be used by DOE to select from among the candidate processes under consideration."

DOE identified eleven criteria to be used in evaluating three alternatives for processing the HLW in the SRS tanks. Based on information presented by representatives from the SRS, the committee concludes that the eleven criteria are reasonable and appropriate and were developed in a transparent way. However, as described in the body of the report, some of the criteria do not appear to be independent of others, and some criteria appear unlikely to discriminate among the process alternatives.

The methodology for using the evaluation criteria is still evolving, and revisions in the weighting factors may be necessary in consideration of the points raised in the body of this report. Preliminary application of the criteria in selection evaluation in three different scoring exercises by DOE has shown little discrimination among the three processes. The committee recommends that the criteria should not be implemented in a way that relies on a single numerical "total score." Rather than averaging and totaling the scores for each criterion, the various criteria should be seen as relevant to different goals and purposes and should be considered individually. Some of the criteria should be used as "go/no go" gates and some should have thresholds for use that demonstrate a given level of difference between the three processes. Also, the committee recommends that DOE should define what are significant differences in the scoring procedure. The committee finds it difficult to see a path forward for this procedure (e.g., adjustment of weighting factors) without these differences being specified. The objective of the evaluation procedure should be to provide adequate information for making a risk-informed decision evaluating the science, technology, operational aspects, time factors, and costs, as well as policy matters not addressed in this evaluation.

Despite limitations in discriminating among the alternatives, the committee recognizes that research and development currently being conducted for the several alternative processes may result in changes in the scores on the eleven criteria. Additionally, the committee finds that the current scoring system for individual criteria can be useful for identifying and following the progress of the research and development program prior to downselection (i.e., a reduction in the number of process alternatives), thereby assisting in determination of where significant further effort is needed for each process.

INTRODUCTION

The National Research Council (NRC) formed a committee, at the request of the U.S. Department of Energy (DOE), to provide an independent technical review of alternatives selected by representatives at the Savannah River Site for processing the high-level radioactive waste salt solutions stored there in tanks. The work of that committee was completed and its findings were reported in *Alternatives for High-Level Waste Salt Processing at the Savannah River Site* (National Research Council, 2000).

After receiving that report, DOE asked the NRC to provide additional advice on the waste processing efforts at the SRS, and a new committee was impaneled to examine the DOE's selection of a process for separating radionuclides from soluble high-level radioactive waste at that site. This newly constituted committee consists of six members of the previous committee, plus three new members whose areas of expertise were needed to address the new charge. The committee was charged with a three-part task:

1) evaluate the adequacy of the criteria that will be used by DOE to select from among the candidate processes under consideration;

2) evaluate the progress and results of the research and development work that is undertaken on these candidate processes; and

3) assess whether the technical uncertainties have been sufficiently resolved to proceed with downsizing the list of candidate processes.

The committee may, at its discretion, also provide comments on the implementation of the selected process.

The purpose of this brief interim report is to address the first of the three tasks.

BACKGROUND

At present three alternative processes remain under consideration for removal of cesium, strontium, and actinides from tank supernate solutions at SRS; namely, small tank precipitation by tetraphenylborate (TPB), ion exchange on crystalline silicotitanate, and caustic side solvent extraction. A brief description of the site's high-level waste program, described in *Alternatives for High-Level Waste Salt Processing at the Savannah River Site* (National Research Council, 2000), is included in this report as Appendix C. A key recommendation in that report was the following:

> *The committee finds that there are potential barriers to implementation of all of the alternative processing options. The committee recommends that Savannah River proceed with a carefully planned and managed research and development (R&D) program for three of the four alternative processing options (small tank precipitation using TPB, crystalline silicotitanate ion exchange, and caustic side solvent extraction, each including monosodium titanate processing for removing strontium and actinides) until enough information is available to make a more defensible and transparent downselection decision. The budget for this R&D should be small relative to the total cost of the processing program, but this investment will be invaluable to overcoming many of the present uncertainties discussed in this report.*

Since that report was issued in August 2000, DOE has funded research and development on the three alternative processes, and significant progress has been made in ameliorating many of the technical uncertainties. DOE noted in its briefings to the committee that tests of all three treatment alternatives have demonstrated their ability to meet functional requirements. On that basis, and with the associated changes in the work programs of the three alternatives and their management, the DOE Tank Focus Area (TFA)[35] has produced downselection criteria.

---

[35] The TFA has the lead responsibility for developing recommendations on both research and development (R&D) direction and the bases for subsequent recommendations on process selection. This group, together with a technical advisory team and a technical working group, interact with a representative of the DOE Office of Environmental

These were presented to the committee at its first meeting on 20-21 November 2000 (Harmon, 2000a, b), and represent the basis for this report.

DOE SELECTION CRITERIA AND GOALS

The TFA and its associated committees and consultants employed systematic and relatively transparent approaches for devising quantifiable evaluation criteria. Using information gathered from other DOE sites and other organizations, they began with twenty criteria and reduced them to the final eleven in an effort to eliminate redundancy and criteria unable to discriminate among the alternatives. The final set of criteria (see Box 1) was approved by the DOE Office of Environmental Management for use in making recommendations on process downselection.

Management responsible for the process development and recommendation for downselection. This representative recommends to the DOE Assistant Secretary for Environmental Management the final determination on the downselection outcome.

> **BOX 1**
> **DOE Criteria for Process Selection at the Savannah River Site**
>
> 1. **Schedule risk**—Risk to the overall project schedule due to high-risk technology issues not being resolved in time to support downselection [to be made in June 2001].
>
> 2. **Project cost reduction potential**—Potential that cost savings in the total project cost can be identified (generally due to flow sheet or equipment arrangement changes that would allow facility footprint reductions).
>
> 3. **Life-cycle costs through decontamination and decommissioning (D&D)**—Total costs to complete all salt processing (including HLW system costs). The focus is on life-cycle costs, but the separate components' total project cost and operating cost also are examined for key differences.
>
> 4. **Technical maturity**—The overall technical maturity of the process flow sheets (including the required strontium and actinide removal steps). EM-50 [DOE Environmental Management Office of Science and Technology] stages of maturity are applied to each unit operation and the results are averaged.
>
> 5. **Implementation confidence**—Amount of relevant process experience (large-scale demonstration or deployment) in the DOE complex and industry for the key equipment used for each cesium removal process. This criterion also includes commercial availability of key components and chemicals.
>
> 6. **Minimize environmental impacts**—Comparative assessment of environmental impacts from secondary waste streams, airborne emissions, and liquid effluents. This criterion also includes the number of Saltstone vaults required for each process.
>
> 7. **Impacts of the interfaces at the Defense Waste Processing Facility (DWPF)**—Cost of implementing the changes (physical modifications) to the interfacing systems and the loss of [glass] canister production caused by outages for equipment installation or transfer line tie-ins.

> 8. **Process simplicity to interfacing systems**—The simplicity of interfacing the alternative cesium removal processes with other high-level waste systems. The simplicity is measured by the number of process unit operations needed for the interface times a difficulty factor for each interface unit operation.
>
> 9. **Levels of safety control mitigation**—Number and type (e.g., passive, active, administrative, preventive, and mitigative) of controls required to maintain the facility in a safe configuration and to protect the worker, public, and environment.
>
> 10. **Maximize process flexibility in throughput**—Capability to operate the process at a higher or lower throughput (turn-up or turn-down) based on the equipment in the current pre-conceptual designs.
>
> 11. **Maximize process simplicity (operability)**—Simplicity of the process as indicated by the number of pieces of equipment (in both the non-radioactive areas and the remotely operated area) and number of jumpers (piping connections) required inside the remotely operated area.

SOURCE: Harmon, 2000a, 2000b (viewgraph on p. 20 entitled "Criteria Weights–Case A"), and H. Harmon, DOE, email communication, January 5, 2001.

In recognition of some commonalties, the eleven criteria for process selection were grouped by the TFA under the set of six goals shown in Box 2. The criteria were used as a measurement for the effectiveness in reaching these goals.

> **BOX 2**
>
> **DOE Goals for Process Selection at SRS**
>
> **Goal 1: Meet schedule** (Criterion 1)
>
> **Goal 2: Minimize cost** (Criteria 2 and 3)
>
> **Goal 3: Minimize technical risk** (Criteria 4 and 5)
>
> **Goal 4: Minimize environmental safety and health impacts** (Criteria 6 and 9)
>
> **Goal 5: Minimize impact to interfaces** (Criteria 7 and 8)
>
> **Goal 6: Maximize process flexibility** (Criteria 10 and 11)

SOURCE: Harmon, 2000b, viewgraph on p. 14 entitled "Criteria Aligned by Goal"

Other possible goals, such as 'minimize tank space requirements' and 'stakeholder acceptance,' were not included by DOE, because they were considered to be integral to the goals listed above or were not considered to be good discriminators among the alternatives.

The TFA employed a series of steps to develop and implement the proposed criteria. In particular, they used several groups of experts to carry out preliminary application of the criteria to evaluation of the three processing alternatives. This preliminary screening was intended to determine if the criteria were capable of distinguishing among the alternatives and to determine to what extent the outcome might depend on the relative weighting assigned to each of the criteria. In conducting this preliminary screening, each alternative was evaluated by the group of experts and assigned an integer score from 1 (worst score) to 5 (best score). The resulting scores were then normalized to generate 'utility values'[36] that ranged from 0 (worst) to 1 (best). Finally, each utility value was multiplied by a weighting factor ranging from 0.03 (low weight) to 0.14 (high weight); the highest weighting was given to technical risk (Criteria 4 and 5). Finally, a total score for each of the alternatives was calculated by summing the eleven individually weighted utility values.

---

[36] The utility value is computed by the formula $u_i = 0.25 (A_i - 1)$, where A_i is the score from 1-5 for criterion i. The total score is then determined by multiplying each utility value by an assigned weighting factor (k_i) and summing the weighted scores. Total Score = $\Sigma (u_i k_I)$

Several preliminary scoring exercises (carried out by various advisory and management groups of the TFA) were reported at the November committee meeting. In all of the exercises the resulting total scores for the three alternative processes all fell within the range of 0.60 to 0.69; in one exercise the identical total score of 0.63 was calculated for all three alternative processes. The actual scoring and weightings were consensus values arrived at in review meetings among the experts following extensive discussion. This consensus represents the informed judgement of these experts. The TFA program plans to reevaluate quarterly the scoring of the alternative processes to take into account the relative progress in the R&D efforts for each alternative. A final downselection decision to one process is scheduled for June 2001.

COMMENTS ON CRITERIA

Criterion 1: Schedule Risk. The time frame for completion of the cleanup activity could readily be modified by subsequent funding or policy decisions or by environmental issues. Hence, while the criterion is generally useful in broad terms, it may not be a significant discriminator among the processes. It might be preferable to employ this criteria on a 'go/no go' basis, in which it would have zero weight unless the calculated risk exceeded the inherent uncertainty.

Criterion 2: Project Cost Reduction Potential. Cost is an important consideration in any project of this magnitude. The costs assigned to the process are likely to be governed largely by the cost of major new facilities, and DOE has carried out extensive cost estimates. These initial estimates did indicate differences between the three processes, but the uncertainty in these estimates is sufficiently large that the projected costs for the three alternatives may be essentially equivalent.[37] Cost reduction would result from divergence from the estimates, so if these have been carried out consistently (i.e., with the same level of conservatism), it is unlikely that the criterion will discriminate among the alternatives. At this early stage, cost estimates are not very accurate, and from a policy standpoint there may be a difference between capital costs and operating costs that makes the current estimate of life cycle costs inadequate as a factor for decision making.

Criterion 3: Life-Cycle Costs Through Decontamination and Decommissioning. The federal budgeting procedure takes place on an annual basis and does not ordinarily include life-cycle costs. In addition, funding from more than one DOE Environmental Management office

---

[37]Kenneth Lang, Department of Energy, oral communication, February 22, 2001.

complicates the financial aspects of the cleanup. Consequently, while life-cycle cost is an important issue, the high uncertainties in DOE cost estimates may limit its value in decision making unless the project is privatized.

Criterion 4: Technical Maturity. This criterion appears to provide reasonable input for the downselection procedure, since the major uncertainties identified by the previous committee (National Research Council, 2000) were in areas of science and technology.

Criterion 5: Implementation Confidence. This criterion evaluates the extent to which a given technology has been demonstrated or deployed at large scale, with higher scores assigned when previously used for processing radioactive materials or used within the DOE complex. This does not appear to be independent of Criterion 4, and if given too large a weighting, could result in double counting.

Criterion 6: Minimize Environmental Impacts. Any process selected for implementation would need to gain the necessary regulatory approval, which will be a clear "yes/no" decision. While it is an appropriate goal for each of the alternative processes to minimize radioactive and chemical emissions and generation of secondary waste, the process to be selected will either meet regulatory approval or it will not. The minimization of waste streams is closely tied to project cost, so this criterion may not be independent of Criteria 2 and 3. Compliance with existing regulations is assumed by DOE and the committee, so comparison of environmental impacts beyond regulatory levels does not represent a relevant and useful discriminator among the three processes.

Criterion 7: Impacts of the Interfaces at the Defense Waste Processing Facility (DWPF). The major focus of this criterion is the process interface with the DWPF, and indirectly with the Saltstone Facility, primarily in terms of number of canisters of vitrified waste to be produced. The DWPF probably represents the most complex and schedule-sensitive operation. Technical modification of these interfaces to allow greater system flexibility would seem to be part of Criteria 4 and 5. In addition, the impact of the interface to DWPF will appear in schedule and costs, so this criterion does not appear to be independent of Criteria 1 through 3.

Criterion 8: Process Simplicity to Interfacing Systems. This is similar to the preceding criterion, and the impact of complexity of the interfaces will appear in schedule and costs.

Criterion 9: Levels of Safety Control Mitigation. As in the case of Criterion 6, regulatory approval will be on a "yes/no" basis, and DOE would only select a process that could be operated safely. The impact of any additional levels of safety control mitigation would appear under cost, so this does not appear to provide discrimination among the alternatives.

Criterion 10: Maximize Process Flexibility in Throughput. This criterion is closely related to several others, including Criterion 1 (schedule), Criteria 4 and 5 (technical), Criterion 7 (interfaces), and Criterion 8 (simplicity and interfaces). While the capability to increase throughput above that of the process design may be desirable for cost factors, such enhancement could have a negative impact on the interfaces with the DWPF and Saltstone operations. Hence, the use of this criterion as a discriminator appears to be in isolation of what should be an integrated system of waste processing. This criterion does not appear to discriminate among the alternatives.

Criterion 11: Maximize Process Simplicity (Operability). The role of simplicity in a process is closely coupled to other factors, including schedule (i.e., lower frequency of process upsets), interfaces with other system processes, and technical risk. However, this criterion may be useful in discriminating among extremes in operability and process complexity, especially where certain operations require very high precision in conditions such as temperature or concentrations.

FINDINGS AND RECOMMENDATIONS

The purpose of this report is to address the first part of the committee's charge: "evaluate the adequacy of the criteria that will be used by DOE to select from among the candidate processes under consideration." The eleven criteria—and the goals under which DOE has grouped them—are reasonable and appropriate and were developed in a transparent way.

> **Finding: The committee finds that DOE's proposed criteria are an acceptable basis for selecting among the candidate processes under consideration; however, as noted in the preceding discussion, some of the criteria do not appear to be independent of others and some criteria appear unlikely to discriminate among the process alternatives.**

The use of the criteria to reach a final decision relies on a methodology that is still evolving. The weighting factors have not yet been decided, and these may need to be adjusted in consideration of the points raised in the previous section about overlap of some criteria or the concepts of go/no go gates and thresholds. In the application of the algorithm to the process alternatives described to the committee there was little discrimination among the alternatives. There was little difference among the total scores, and the ranking appeared to be dependent upon the weighting factors employed. This raises the question of whether the algorithm is capable of providing adequate discrimination among the alternatives. Is it possible that high scores for certain criteria could obscure serious problems in other criteria?

> **Recommendation: The committee recommends that the criteria should not be implemented in a way that relies on a single numerical "total score." Rather than averaging and totaling the scores for each criterion, the various criteria should be seen as relevant to different goals and purposes and should be considered individually. Some of the criteria should be used as "go/no go" gates and some should have thresholds for use.**

Despite limitations in discriminating among the alternatives, the committee recognizes that R&D progress for the several alternative processes may result in changes in the respective scores on the eleven criteria.

> **Finding: The committee finds that the current scoring system for individual criteria can be useful for identifying and following the progress of research and development program prior to a final downselection. This could assist in determining where significant further effort is needed for each process.**

The final selection of a process for treating the SRS high-level waste will be a management decision. The final decision rests with the Assistant Secretary of Energy for Environmental Management and will be made on the basis of documentation related to the eleven criteria discussed here. The committee believes that the proposed criteria can provide adequate information for making a risk-informed decision evaluating the science, technology, operational aspects, time factors, costs, and policy matters. As indicated in the preceding comments on the criteria, some issues—for, example, life-cycle costs—do not match well with the federal procedure for allocating funds. This would not be the case for a privatized operation, and if a contractor were responsible for costs it might be necessary for them to be involved formally in the decision-making procedure.

REFERENCES CITED

Harmon, H.D. 2000a (September 15). Viewgraphs entitled *Salt Processing Project: Bases for Scoring of Alternative Cesium Removal Processes on August 14-15, 2000; Recorded by The Tanks Focus Area for the Technical Working Group.* U.S. Department of Energy Tanks Focus Area. 38pp.

Harmon, H.D. 2000b (November 20). Viewgraphs entitled *Salt Processing Project Down-Selection Criteria.* U.S. Department of Energy Tanks Focus Area. 24pp.

National Research Council. 2000. Alternatives for High-Level Waste Salt Processing at the Savannah River Site. Washington, D.C.: National Academy Press, 142pp.

APPENDIX A

BIOGRAPHICAL SKETCHES OF COMMITTEE MEMBERS

MILTON LEVENSON (*Chair*) is a chemical engineer with over 48 years of experience in nuclear energy and related fields. His technical experience includes work in nuclear safety, fuel cycle, water reactor technology, advanced reactor technology, remote control technology, and sodium reactor technology. His professional experience includes positions at Oak Ridge National Laboratory (research and operations), Argonne National Laboratory, the Electric Power Research Institute (first director of nuclear power), and Bechtel (last position was vice-president of Bechtel International). Mr. Levenson is the past president of the American Nuclear Society and a fellow of the American Nuclear Society and the American Institute of Chemical Engineers. He is the author of over 150 publications and holds three U.S. patents. He was elected to the National Academy of Engineering in 1976. Mr. Levenson has served on many National Research Council committees, and in 1998 served as principal investigator for the Board on Radioactive Waste Management project on aluminum spent fuel.

GREGORY R. CHOPPIN (*Vice-Chair*) is the R.O. Lawton Distinguished Professor of Chemistry at Florida State University. His research interests include nuclear chemistry, physical chemistry of actinides and lanthanides, environmental behavior of actinides, chemistry of the f-elements, separation science of the f-elements, and concentrated electrolyte solutions. While at Lawrence Radiation Laboratory, University of California, Berkeley, he participated in the discovery of mendelevium, element 101. Dr. Choppin's research interests have been recognized by the American Chemical Society with its Award in Nuclear Chemistry and the Southern Chemist Award, the Manufacturing Chemists Award in Chemical Education, and a with a Presidential Citation Award by the American Nuclear Society. He has served on numerous NRC committees, is currently a member of the Board on Radioactive Waste Management, and recently completed a 6-year term as a member of the Board on Chemical Sciences and Technology.

JOHN E. BERCAW is the Centennial Professor of Chemistry at the California Institute of Technology. Dr. Bercaw is an expert in organometallic chemistry. His research interests include synthetic, structural, and mechanistic organotransition metal chemistry, compounds of early transition metals, and hydrolyzation of alkanes by simple platinum halides in aqueous solutions. Dr. Bercaw is a former chair and executive committee member of the American Chemical Society's Inorganic Chemistry Division. He is a fellow of the American Association for the Advancement of Science and of the American Academy of Arts and Sciences. His work has been recognized by the American Chemical Society with its Award in Pure Chemistry, the George A. Olah Award in Hydrocarbon or Petroleum Chemistry, and the Award for Distinguished Service in the Advancement of Inorganic Chemistry. Dr. Bercaw was elected to the National Academy of Sciences in 1990.

DARYLE H. BUSCH is the Roy A. Roberts Distinguished Professor of chemistry at the University of Kansas. His research, which fathered synthetic macrocyclic ligand chemistry and created the molecular template effect, is presently focused on homogeneous catalysis, bioinorganic chemistry, and orderly molecular entanglements. He is a recipient of the American Chemical Society's Award for Distinguished Service in Inorganic Chemistry and its Award for Research in Inorganic Chemistry. Recently Dr. Busch received the International Izatt-Christensen Award for Research in Macrocyclic Chemistry and the University of Kansas's Louis Byrd Graduate Educator Award. Dr. Busch was elected president of the American Chemical Society in 2000.

JAMES H. ESPENSON is Distinguished Professor of Chemistry at Iowa State University and program director of molecular processes at DOE's Ames Laboratory. He has received the John A. Wilkinson award for excellence in teaching, an award from the Alfred P. Sloan Foundation, and is a fellow of the American Association for the Advancement of Science. He has served as a member of the executive committee and as a councilor for the American Chemical Society's Division of Inorganic Chemistry. Espenson studies transition metal complexes as catalysts for chemical reactions (including oxidation-reduction reactions), as participants in atom-transfer mechanisms, as reagents in new reactions, and as templates for coordination phenomena. His research has focused on oxo- and thio-complexes of rhenium in high oxidation states.

GEORGE E. KELLER II, since retiring as a senior corporate research fellow from the Union Carbide Corporation in 1997, has been active in economic-development enterprises and consulting. He is also an adjunct professor of chemical engineering at two universities. His technical expertise lies in separation processes, reaction engineering and catalysis, energy use minimization, and new process configurations. Dr. Keller has 35 publications and 21 co-held patents, and has given invited lectures in many universities, technical meetings, and companies around the world. He is the recipient of four national awards for technical excellence: three from the American Institute of Chemical Engineers and the Chemical

Pioneer Award from the American Institute of Chemists. He was elected to the National Academy of Engineering in 1988 and presently serves as a member of the Board on Chemical Sciences and Technology of the National Research Council.

THEODORE A. KOCH is currently a DuPont fellow (the highest professional title in the company); he is also an adjunct professor of chemical engineering at the University of Delaware. He has spent his entire career developing chemical processes and bringing them from the benchtop to commercial reality. He holds 29 patents and has authored 9 journal articles and 1 book. He is a member of the Catalysis Club of Philadelphia (former program chair and president), the North American Catalysis Society, and the American Institute of Chemical Engineers. Dr. Koch received the Award for Excellence in Catalytic Science and Technology from the Catalysis Club of Philadelphia and the Lavoisier Award for Technical Excellence from the E.I. du Pont de Nemours and Company.

ALFRED P. SATTELBERGER is the director of the Chemistry Division at Los Alamos National Laboratory. Dr. Sattelberger's research interests include actinide science, technetium coordination and organometallic chemistry, and metal-metal multiple bonding. Prior to his current position Dr. Sattelberger held a professorship at the University of Michigan. He is a past member of the executive committee of the Inorganic Chemistry Division of the American Chemical Society, and serves on the board of directors for the Inorganic Synthesis Corporation and the Los Alamos National Laboratory Foundation. He served as a reviewer on the FY 1996 general inorganic chemistry Environmental Management Science Program merit review panel and on the National Research Council's Committee on Building an Effective EM Science Program.

MARTIN J. STEINDLER'S last position was as director of the Chemical Technology Division at Argonne National Laboratory. His expertise is in the fields of nuclear fuel cycle and associated chemistry, engineering, and safety, with emphasis on fission products and actinides. He also has experience in the structure and management of research, development, and deployment organizations and activities. Dr. Steindler has been a consultant to the Atomic Energy Commission, the Energy Research and Development Agency, and various Department of Energy laboratories. He chaired both the Materials Review Board for the DOE Office of Civilian Radioactive Waste Management and the U.S. Nuclear Regulatory Commission Advisory Committee on Nuclear Waste. Dr.

Steindler has served on several National Research Council committees, and currently serves on the Board on Radioactive Waste Management.

APPENDIX B

LETTERS OF REQUEST FOR THIS STUDY

Department of Energy
Washington, DC 20585
April 13, 2000

Dr. Kevin D. Crowley
Director
Board on Radioactive Waste Management
National Research Council
2001 Wisconsin Avenue, N. W. Washington, DC 20007

Dear Dr. Crowley:

I would like to take this opportunity to thank you and your Committee Members for your extraordinary effort providing the Department with an independent technical review of alternatives for processing the high-level radioactive waste salt solutions at the Savannah River Site. We agree with your interim comments noting that additional research and development is required for each option, and we are proceeding with addressing your comments in our research and development plans for fiscal years 2000 and 2001. I am looking forward to receiving your final report this month so that we can make adjustments in the current plans if needed.

I believe that the complexity of the salt processing technology alternatives warrants your continued involvement in our continuing research and development
efforts. Therefore, I would like to request that you and your Committee continue to support the Department throughout the next year by providing us with your independent review of each technology road map, and the selection criteria.

Dr. Huntoon has tasked me, as the Deputy Assistant Secretary for the Office of Project Completion, to provide the leadership and program management for technology development and selection of a preferred treatment alternative. I am working closely with the Office of Science and Technology, as well as the DOE-Savannah River Operations Office, to make sure that this effort is adequately supported. An Action Plan has been prepared, and is enclosed, which provides details of the roles and responsibilities for the project. I will be providing the Assistant Secretary for Environmental Management with quarterly progress reviews on each of the technology activities throughout the ensuing months, and I propose that we follow those reviews with a briefing to your Committee to keep you abreast of the salt processing project's progress. Of course, additional briefings, meetings, and documentation will be made available to the Committee as you deem necessary to support your review.

Based on the current schedule, we would be seeking your Committee's review of the items identified above in early summer 2000. Due to the short time available between now and the anticipated time we require your support, may I suggest that you utilize your existing Committee to expedite matters.

Mr. Ken Lang of my staff will be contacting you directly to coordinate the details.
Mr. Lang can be reached at (301) 903- 7453.

Thank you in advance for your continued support of DOE. I look forward to
working with you in this endeavor.

Sincerely,

Mark W. Frei
Deputy Assistant Secretary
for Project Completion
Office of Project Completion

Enclosure

cc:
M. Gilbertson, EM-52
K. Picha, EM-22
G. Rudy, DOE-SR
K. Gerdes, EM-54
B. Spader, DOE-SR
J. Case, DOE-ID

Department of Energy
Washington, DC 20585
June 15, 2000

Dr. Kevin D. Crowley
Director
Board on Radioactive Waste Management
National Research Council
2001 Wisconsin Avenue, NW
Washington, D.C. 20007

Dear Dr. Crowley:

Thank you for your May 16, 2000, letter responding to my request that the National Research Council continue its support of the Department's high-level waste salt processing alternatives at the Savannah River Site.

I am pleased that you would like to continue to provide technical assistance to the Department throughout the planned research and development phase of this project, pending approval of the Board on Radioactive Waste Management and the National Research Council Governing Board.

Your proposals to (I) comment on the criteria that will be used to select a processing alternative; (2) evaluate the results of the research and development work that is undertaken on the candidate processing alternatives; and (3) provide the Department with an assessment of whether the technical uncertainties have been sufficiently resolved to proceed with downsizing the list of alternatives will meet our needs throughout the remaining research and development period.
I found the interim report you provided on your current evaluation to be particularly
useful in planning the research and development now underway, and I am confident that an interim report for this phase of the study will be valuable in the selection of alternative processing technologies.

Mr. Kenneth Lang of my staff is available to support you and the committee for this review. Mr. Lang can be reached at (301) 903- 7453.

Thank you for your continued support of DOE.

Sincerely,

Mark W. Frei
Deputy Assistant Secretary
for Project Completion
Office of Environmental Management

cc:
M. Gilbertson, EM-52
K. Picha, EM-22
G. Rudy, DOE-SR
K. Gerdes, EM-54
G. Boyd, EM-50
B. Spader, DOE-SR
J. Case, DOE-ID

APPENDIX C
HIGH-LEVEL WASTE AT THE SAVANNAH RIVER SITE

During and immediately following the Second World War, the U.S. Government established large industrial complexes at several sites across the United States to develop, manufacture, and test nuclear weapons. One of these complexes was established in 1950 at the Savannah River Site (SRS) to produce isotopes, mainly plutonium and tritium, for defense purposes. The site is located adjacent to the Savannah River near the Georgia-South Carolina border and the city of Augusta, Georgia, and comprises an area of about 800 square kilometers (~300 square miles).

The SRS was host to an extensive complex of facilities that included fuel and target fabrication plants, nuclear reactors, chemical processing plants, underground storage tanks, and waste processing and immobilization facilities. Plutonium and tritium were produced by irradiating specially prepared metal targets in the nuclear reactors at the site. After irradiation, the targets were transferred to canyon facilities, where they were processed chemically to recover these radionuclides. This processing resulted in the production of large amounts of highly radioactive liquid waste, known as *high-level waste* (HLW), that, after treatment with caustic, is being stored in two underground tank farms at the site.

TANK WASTE PROCESSING

DOE has the responsibility for waste management at SRS and has implemented a program to stabilize this HLW and close the tank farms. The tank waste processing system at SRS comprises the major components; (a) waste concentration and storage, (b) radionuclide immobilization, (c) extended sludge processing, (d) salt processing, and (e) salt disposal.

Waste Concentration and Storage

The high-level waste resulting from operations in the chemical processing canyons is currently being stored in 48 underground carbon-steel tanks. The tanks range in size from about 3 million to 5 million liters (750,000 to 1.3 million gallons). The HLW was made alkaline with sodium hydroxide (NaOH) and formed a caustic sludge before being transferred to the tanks to reduce corrosion of the carbon steel primary containment. Consequently, the waste has a high pH (>14) and a high salt (especially sodium) content.

Approximately 400 million liters (100 million gallons) of HLW were produced at SRS since operations began in the 1950s, but this volume has been reduced to about 130 million liters (34 million gallons) by removal of excess water through evaporator processing operations. About 10 percent of the waste by volume is in the form of a water-insoluble precipitate, or *sludge*, that contains most of the actinides (i.e., uranium as well as transuranic elements) and strontium-90. This sludge was formed by natural settling and by precipitation when NaOH was added to the waste. The remaining waste consists of solid sodium salts (*saltcake*) and an aqueous solution (saturated with sodium salts) called supernate (which contains approximately 95 percent of the cesium in the tank waste, as well as minor amounts of actinides). The saltcake, produced by crystallization after the alkaline waste was processed through evaporators to reduce the volume of material, will dissolve when additional water is added during waste processing. The saltcake and sludge contain substantial quantities of supernate within their mass; this interstitial supernate corresponds to about half of the total supernate in the tanks.

Radionuclide Immobilization

The Defense Waste Processing Facility (DWPF) was constructed to immobilize radioactive waste in borosilicate glass for eventual shipment to and disposal in a geological repository. The glass-making process is referred to as *vitrification*. This glass is produced by combining the processed HLW (the processing operations are discussed below) with specially formulated glass frit and melting the mixture at about 1150 °C. The molten glass is then poured into cylindrical stainless steel canisters, allowed to cool, and sealed. The DWPF canisters are about 60 centimeters (2 feet) in diameter and about 300 centimeters (10 feet) in length and contain about 1,800 kilograms (4,000 pounds) of glass. About 700 canisters have been produced to date[38], and SRS estimates that a total of about 6,000 canisters would be produced by 2026, when the tank waste processing program is planned to be completed. These canisters are to be stored at the site until a permanent geological repository is opened and ready to receive them.

Extended Sludge Processing

Extended sludge processing is being used to prepare the sludge portion of the tank waste for processing into glass. The sludge is removed

---

[38]Since this appendix was originally published, over 300 additional canisters have been produced.

from the tanks by hydraulic slurrying and washed to remove aluminum and soluble salts, both of which can interfere with the glass-making process. The washed sludge is transferred to the DWPF for further processing before being incorporated into glass. Sludge processing would result in immobilization in glass of nearly all of the strontium and actinides from the tanks.

Salt Processing

Salt processing would be used to remove much of the radionuclides from the HLW salt for eventual vitrification. The salt is to be redissolved and transferred out of the tanks. It would then be mixed with a sorbent to remove any remaining actinides (mainly uranium and plutonium) and strontium. The currently planned sorbent is monosodium titanate (MST). The solution will then be subjected to another (and as-yet undetermined) process to remove cesium. The separated actinides, strontium, and cesium would be washed to remove soluble salts and sent to the DWPF for immobilization.

Salt Disposal.

A variety of secondary waste streams are formed during the processing operations described above. Some of these waste streams are recycled back to the tanks, some are recycled within the various processing operations, and yet other wastes are treated and stabilized for burial. Most notably, the "decontaminated" salt supernate (i.e., the solutions remaining after actinide, strontium, and cesium removal) would be disposed of onsite in a waste form known as *Saltstone*. The residual solutions are classified as "incidental waste" from the processing of HLW. Saltstone is created by mixing the residual salt solutions with fly ash, slag, and Portland cement to create a grout slurry. This slurry is then poured into concrete vaults, where it cures (solidifies) and is eventually covered with soil. The Saltstone contains small quantities of some radionuclides.

CESIUM REMOVAL PROBLEM

As noted above, SRS planned to remove actinides, strontium, and cesium from the salt solutions in two processing steps. First, actinides and strontium were to be removed by mixing the salt solutions with MST, resulting in the sorption of actinides and strontium. The product of this reaction could be removed from the salt solutions by filtration for subsequent processing and immobilization. Subsequently, the removal of cesium from the salt solutions would be accomplished by a yet-to-be-chosen process from among precipitation, ion exchange or solvent extraction processes.

In the late 1970s and the 1980s, SRS developed a process for removing cesium from salt solutions through a precipitation reaction involving sodium tetraphenylborate (NaTPB) and cesium to form cesium TPB (CsTPB): SRS refers to this process as "In-Tank Precipitation." The NaTPB was to be added directly to a large waste tank to produce a cesium-bearing precipitate. SRS undertook an ITP pilot project in 1983 to demonstrate proof of principle. The process removed cesium from the salt solution, but it also resulted in the generation of flammable benzene from radiolytic reactions and possibly from catalytic reactions with trace metals in the waste. In September 1995, SRS initiated ITP processing operations in a tank that contained about 1.7 million liters (450,000 gallons) of salt solutions. The operations were halted after about 3 months because of higher-than-expected rates of benzene generation. SRS staff then initiated a research program to develop a better understanding of the mechanisms of benzene generation and release. They also considered possible design changes to handle the benzene during processing operations and catalyst poisoning strategies.

In 1996, the Defense Nuclear Facility Safety Board (DFNSB) issued Recommendation 96-1, urging DOE to halt all further testing and to begin an investigative effort to understand the mechanisms of benzene formation and release. Investigations by SRS in 1997 uncovered the possible role of metal catalysts in the benzene formation process. SRS concluded, however, that both safety and production requirements could not be met, which led to the suspension of operations altogether in early 1998. At the time of suspension, SRS had spent almost a half billion dollars to develop and implement the ITP process. In March 1998, Westinghouse Savannah River Company (WSRC) formed a systems engineering team to identify alternatives to the ITP process for separating cesium. This team began by undertaking a literature and patent screening procedure to identify currently known processes, followed by a system of analyses by panels of experts to reduce the number of alternative processes to four.

Strontium/Actinide Removal by MST

In all four of the final candidate processes for cesium separation, prior removal of strontium and actinides is viewed by SRS as a requisite process. At present, the use of MST is the method of choice. Some technical uncertainties remain to be resolved, of which the major ones are the kinetics of sorption on MST and the amount of titanate acceptable for proper quality of the vitrified waste form.

Tetraphenylborate Precipitation Process

The ITP developed by WSRC removes cesium from HLW supernates by precipitation with tetraphenylborate ion, $[B(C_6H_5)_4]^-$ (TPB). Sodium TPB is a reagent used for analyzing for the potassium ion based on the insolubility of potassium TPB (KTPB). The 200-fold lower solubility of cesium TPB (CsTPB) can provide decontamination factors (DF) from the salt as high as 10^5 to 10^6 and the mixed CsTPB/KTPB precipitate is typically in a form that is easily filtered. On the average, the SRS HLW in the waste tanks contains sodium ions (approximately 5 molar), potassium ions (approximately 0.03 molar), and cesium ions (approximately 0.00025 molar).

HLW treatment, including the removal of cesium-137, involves separation of selected radioactive components and their subsequent immobilization in a borosilicate glass at the DWPF. To prevent organic material from being fed to the DWPF melters, the CsTPB/KTPB precipitate must be treated to remove more than 90 percent of the phenyl (C_6H_5) groups bound to the boron. Thus, a precipitate hydrolysis process (PHP) was developed to hydrolyze the TPB using formic acid in the presence of a copper catalyst. The hydrolysis products are benzene, which is removed by evaporation and incineration, and an aqueous solution containing $^{137}Cs^+$, $B(OH)_3$, and K^+ ions. An attractive feature of TPB is its susceptibility to catalytic decomposition downstream.

Crystalline Silicotitanate Ion Exchange

Ion exchange has been in commercial use for over 100 years to remove ions from aqueous solutions, e.g., to make deionized water. In most applications the separated ions are *eluted* from the ion exchange material, e.g., using a dilute acid, the eluted ions are concentrated, and the ion exchanger is reused over and over. Although this technology is well established, ion exchange for cesium removal from high-level waste at SRS and other DOE sites poses challenges. The ion exchange material must withstand both high alkalinity and high radiation fields and must be very selective for cesium in the presence of much higher concentrations of the chemically related sodium and potassium ions. A promising material for use by SRS to remove cesium is crystalline silicotitanate (CST), developed by Sandia National Laboratory and Texas A&M University, based on work performed on amorphous hydrous titanium oxide in the 1960s and 1970s at Sandia. CST has received considerable attention because of its promise as an ion exchange material for nuclear waste applications. The material has a high selectivity for Cs^+ in salt solutions over a large portion of the pH range from acidic to basic solution, and exhibits high stability to radiation as well. CST is also unusual in that cesium is difficult to remove from the material (i.e., it is nonelutable and the CST cannot be reused). As a result, CST must be incorporated into the HLW stream along with the radionuclides, and the stability of borosilicate glass with higher concentrations of titanium is an issue that must be addressed.

Caustic Side Solvent Extraction

A typical solvent extraction process includes four steps. First, a feed stream is contacted with a solvent that is virtually insoluble in the stream. During this contact, one or more components of the stream transfer to the solvent, while other components do not. The loaded solvent, scrubbed to remove minor contaminants and leaving relatively clean solvent plus the component(s) to be finally recovered, is sent to a stripping operation where the component(s) to be recovered is removed. The stripped solvent may then go to a solvent-recovery step, in which it is cleaned prior to returning to the first step. In such a process, very high removals of extracted components often can be attained.

Solvent extraction has had a long history of successful use in the nuclear industry for such operations as spent fuel reprocessing and plutonium recovery. This history includes long periods of time in which solvents of various organic species have been exposed to high-radiation fields without experiencing catastrophic degradation rates. Solvent extraction operations usually consist of selectively transferring components from an aqueous, acidic stream into the organic stream. A second aqueous stream of somewhat different composition is often used to strip the solvent and concentrate the extract. For the SRS application, the solvent extraction process must remove approximately 99.998 percent of the cesium (a decontamination factor, or DF, of 50,000) from an aqueous, tank-waste feed stream. The raffinate aqueous stream, thus purified of cesium, would be sent to the SRS Saltstone Facility, and the extract, concentrated in cesium by about an order of magnitude is sent to the DWPF.

Direct Disposal in Grout

Direct disposal of the tank waste following removal of strontium and actinides is very similar to the Saltstone process that was to have been used to dispose of the salt solutions from ITP operations as low-level incidental waste. Although it is a rather mature technology and has already been demonstrated at the site for less radioactive salt solutions, the degree of retention of cesium may not satisfy regulatory requirements.

APPENDIX C

Biographical Sketches of Committee Members

MILTON LEVENSON (*Chair*) is a chemical engineer with more than 48 years of experience in nuclear energy and related fields. His technical experience includes work in nuclear safety, fuel cycle, water reactor technology, advanced reactor technology, remote control technology, and sodium reactor technology. His professional experience includes positions at Oak Ridge National Laboratory (research and operations), Argonne National Laboratory, the Electric Power Research Institute (first director of nuclear power), and Bechtel (last position was vice-president of Bechtel International). Mr. Levenson is the past president of the American Nuclear Society and a fellow of the American Nuclear Society and the American Institute of Chemical Engineers. He is the author of more than 150 publications and holds three U.S. patents. He was elected to the National Academy of Engineering in 1976. Mr. Levenson has served on many National Research Council committees, and he served as principal investigator for the Board on Radioactive Waste Management project on aluminum spent fuel in 1998.

GREGORY R. CHOPPIN (*Vice-Chair*) is the R.O. Lawton Distinguished Professor of Chemistry at Florida State University. His research interests include nuclear chemistry, physical chemistry of actinides and lanthanides, environmental behavior of actinides, chemistry of the f-elements, separation science of the f-elements, and concentrated electrolyte solutions. While at Lawrence Radiation Laboratory, University of California, Berkeley, he participated in the discovery of mendelevium, element 101. Dr. Choppin's research interests have been recognized by the American Chemical Society's Award in Nuclear Chemistry and the Southern Chemist Award, the Manufacturing Chemists Award in Chemical Education, and a Presidential Citation Award of the American Nuclear Society. He has served on numerous NRC committees, is currently a member of the BRWM, and recently completed a 6-year term as a member of the Board on Chemical Sciences and Technology.

JOHN E. BERCAW is the Centennial Professor of Chemistry at the California Institute of Technology. Dr. Bercaw is an expert in organometallic chemistry. His research interests include synthetic, structural, and mechanistic organotransition metal chemistry, compounds of early transition metals, and hydroxylation of alkanes by simple platinum halides in aqueous solutions. Dr. Bercaw is a former chair and executive committee member of the American Chemical Society's Inorganic Chemistry Division. He is a fellow of the American Association for the Advancement of Science and a fellow of the American Academy of Arts and Sciences. His work has been recognized with the American Chemical Society's Award in Pure Chemistry, the George A. Olah Award in Hydrocarbon or Petroleum Chemistry, and the American Chemical Society's Award for Distinguished Service in the Advancement of Inorganic Chemistry. Dr. Bercaw was elected to the National Academy of Sciences in 1990.

DARYLE H. BUSCH is the Roy A. Roberts Distinguished Professor of Chemistry at the University of Kansas. His research fathered synthetic macrocyclic ligand chemistry and created the molecular template effect, and is presently focused on homogeneous catalysis, bioinorganic chemistry, and orderly molecular entanglements. He is a recipient of American Chemical Society Awards for Distinguished Service in Inorganic Chemistry and for Research in Inorganic Chemistry. Recently, Dr. Busch received the International Izatt-Christensen Award for Research in Macrocyclic Chemistry and the University of Kansas' Louis Byrd Graduate Educator Award. Dr. Busch was elected president of the American Chemical Society in 2000.

JAMES H. ESPENSON is Distinguished Professor of Chemistry at Iowa State University, and program director of molecular processes at the Department of Energy's Ames Laboratory. He has received the John A. Wilkinson award for excellence in teaching, and an award from the Alfred P. Sloan Foundation, and he is a fellow of the American Association for the Advancement of Science. He has served as a member of the executive committee and as a councilor for the American Chemical Society Division of Inorganic Chemistry. Espenson studies transition metal complexes as catalysts for chemical reactions (including oxidation-reduction reactions); as participants in atom-transfer mechanisms; as reagents in new reactions; and as templates for coordination phenomena. His research has focused on oxo and thio complexes of rhenium in high oxidation states.

GEORGE E. KELLER II, since retiring as a senior corporate research fellow from the Union Carbide Corporation in 1997, has been active in economic development enterprises and consulting. He is also an adjunct professor of chemical engineering at two universities. His technical expertise lies in separation processes, reaction engineering and catalysis, energy use minimization, and new process configurations. Dr. Keller has 35 publications and 21 co-held patents, and has given invited lectures at many universities, technical meetings, and companies around the world. He is the recipient of four national awards for technical excellence—three from the American Institute of Chemical Engineers and the Chemical Pioneer Award from the American Institute of Chemists. He was elected to the National Academy of Engineering in 1988 and presently serves as a member of the Board on Chemical Sciences and Technology of the National Research Council.

THEODORE A. KOCH is currently a DuPont Co. fellow (the highest professional title in the company); in addition, he is an adjunct professor of chemical engineering at the University of Delaware. He has spent his entire career developing chemical processes and bringing them from the benchtop through commercial reality. He holds 29 patents and has authored 9 journal articles and 1 book. He is a member of the Catalysis Club of Philadelphia (former program chair and president), the North American Catalysis Society, and the American Institute of Chemical Engineers. Dr. Koch received the Award for Excellence in Catalytic Science and Technology from the Catalysis Club of Philadelphia and the Lavoisier Award for Technical Excellence from the DuPont Co.

ALFRED P. SATTELBERGER is director of the Chemistry Division at Los Alamos National Laboratory. Dr. Sattelberger's research interests include actinide science, technetium coordination and organometallic chemistry, and metal-metal multiple bonding. Prior to his current position, Dr. Sattelberger held a professorship at the University of Michigan. He is a past member of the executive committee of the Inorganic Chemistry Division of the American Chemical Society and serves on the board of directors for the Inorganic Synthesis Corporation and the Los Alamos National Laboratory Foundation. He served as a reviewer on the fiscal year 96 general inorganic chemistry Environmental Management Science Program merit review panel and on the NRC Committee on Building an Effective EM Science Program.

MARTIN J. STEINDLER'S last position was as director of the Chemical Technology Division at Argonne National Laboratory. His expertise is in the fields of nuclear fuel cycle and associated chemistry, engineering, and safety with emphasis on fission products and actinides. In addition, he has experience in the structure and management of research, development, and deployment organizations and activities. During his career, Dr. Steindler has been a consultant to the Atomic Energy Commission, the Energy Research and Development Agency, and various

Department of Energy laboratories. He chaired the Materials Review Board for the U.S. Department of Energy Office of Civilian Radioactive Waste Management and the U.S. Nuclear Regulatory Commission Advisory Committee on Nuclear Waste. Dr. Steindler has served on several National Research Council committees and currently serves on the BRWM.

APPENDIX D

Information-Gathering Meetings

PRESENTATIONS GIVENT DURING SECOND COMMITTEE MEETING
February 13 - February 14, 2001, Augusta, Georgia

Salt Processing Technology Development Overview (Harry Harmon, SRS)

Alpha and Strontium Removal: Program Overview (Samuel Fink, SRS)

Alpha and Strontium Removal: Sorbent Studies (David Hobbs, SRS)

Alpha and Strontium Removal: Solid-Liquid Separation Studies (Michael Poirier, SRS)

Caustic Side Solvent Extraction: Overview (Major Thompson, SRS)

CSSX: Flowsheet Test Results (Ralph Leonard, ANL)

CSSX: Solvent Chemical and Physical Properties (Bruce Moyer, ORNL)

CSSX: Solvent Chemical and Thermal Stability (Bruce Moyer, ORNL)

CSSX: Solvent Radiolytic Stability (Leon Klatt, ORNL)

CSSX: Radioactive Waste Tests (Doug Walker, SRTC)

CSSX: Solvent Preparation and Commercialization (Peter Bonnesen, ORNL)

Crystalline Silicotitanate Nonelutable Ion Exchange: Overview (Dennis Wester, PNNL)

CST: Gas Generation (Dennis Wester, PNNL)

CST: Chemical and Thermal Stability (Doug Walker, SRTC)

CST: Chemical and Thermal Stability Studies of Cesium-Loaded IE-911 (Liyu Li, PNNL)

CST: Pretreatment Technologies for IE-911 Development (Jim Krumhansl, SNL)

CST: Performance of IE-911: Characterization of As-Received, NaOH-Treated, and Simulant-Treated CST (May Nyman, SNL)

CST: Effect of Organic Impurities and Minor Components on Cesium Sorption (Fernando Fondeur, SRTC)

CST: Sorbent Handling and Sampling (Frank Smith, SRTC)

Small Tank Tetraphenylborate Precipitation: Overview (Joe Walker ORNL)

STTP: Tetraphenylborate Decomposition and Catalyst Studies (Mark Barnes, SRTC)

STTP: NMR Studies for Catalyst Understanding (Peter Bonneson, ORNL)

STTP: Characterization of Palladium and Mercury after Reaction with Dissolved Tetraphenylboron (Martine Duff, University of Georgia)

STTP: Catalyst Development Overview (Jim Boncella, University of Florida)

STTP: Antifoam Development and Testing (Dan Lambert, SRTC)

STTP: Real Waste Tests (Mark Barnes, SRTC)

STTP: 20-Liter Continuous Stirred Tank Reactor Studies (Doug Lee, ORNL)

PRESENTATIONS GIVEN DURING THE THIRD COMMITTEE MEETING
March 26-27, 2001, Washington, D.C.

High-Level Waste Characterization (Joe Carter, SRS)

Salt Processing Project Technology Development Update (Harry Harmon, SRS)

APPENDIX E

Acronyms and Abbreviations

ANL	Argonne National Laboratory
Cs	cesium
CsTPB	cesium tetraphenylborate
CST	crystalline silicotitanate
CSSX	caustic side solvent extraction
DF	decontamination factor
DOE	United States Department of Energy
EW	aqueous strip effluent
DWPF	Defense Waste Processing Facility
EXAPS	extended x-ray adsorption fine structure spectroscopy
fcc	face-centered cubic
FS	feed simulant
$HgPh_2$	diphenylmercury
HLW	high-level waste
ITP	in-tank precipitation
MST	monosodium titanate
NaTPB	sodium tetraphenylborate
NMR	nuclear magnetic resonance
Np	neptunium
NRC	National Research Council
ORNL	Oak Ridge National Laboratory
PNNL	Pacific Northwest National Laboratory
Pu	plutonium
R&D	research and development
SNL	Sandia National Laboratory
Sr	strontium
SRS	Savannah River Site
SRTC	Savannah River Technology Center
ST	sodium nonatitanate
STTP	small tank tetraphenylborate precipitation
TEM	transmission electron spectroscopy
TPB	tetraphenylborate ion $[B(C_6H_5)_4]^-$
U	uranium
WSRC	Westinghouse Savannah River Company